Herd Wired

—·—

In Pursuit of Discovery

Kerry M. Thomas

Beyond The Hoof Bookworks

Edited By: Juliet Murphy

Creative Director: Daphne Thomas

Copyright ©2025 by Kerry M. Thomas

All rights reserved.

No portion of this book may be reproduced in any form without written permission from the publisher or author, except as permitted by U.S. copyright law.

Contents

Dedication	1
Foreword	2
AUTHOR'S NOTE	4
INTRODUCTION	5
Trailhead: Understanding Herd Structure	9
PART ONE	15
Trail Marker One: The Emotional Operating System	16
Trail Marker Two: Foundations of Thought	23
Trail Marker Three: Mapping the Invisible	31
Trail Marker Four: When Connection Breaks	40
Trail Marker Five: The Psychology of Herd Structure	49
Trail Marker Six: Fragile Nature	55
Trail Marker Seven: Environmental Design	65
PART TWO	76
Trail Marker Eight: Processing Pressure	78
Trail Marker Nine: Beyond Body Language	86
Trail Marker Ten: The Emotional Map	93
Trail Marker Eleven: Interruption vs. Fracture	102

Trail Marker Twelve: Behavioral Landscape	110
Trail Marker Thirteen: The Gift of Going Slowly	116
Trail Marker Fourteen: The Psychology of Place	121
Trail Marker Fifteen: Displacement Behaviors & Silent Signals	125
Trail Marker Sixteen: The Independent Nature	132
Trail Marker Seventeen: The North Star Never Moves	137
Glossary of Terms	143
A Note From The Field	149
About the Author	151

Dedication

To Daphne, for always believing.

Foreword

I can still recall the evening Dr. Manon Koning, a Dutch veterinarian, called me: *"You have to see this; it fits exactly what you do."* With infectious enthusiasm she explained how she had discovered Kerry M. Thomas online and how much his insights had helped her and her horse. "Can't you bring Kerry to the Netherlands?" She asked. And so we did.

For more than twenty years Kerry has crossed the dusty plains of the American West, studying mustangs at their most vulnerable and authentic moments. Because he is profoundly color blind, he cannot distinguish horses by their coat; this forced him to focus on movement, energy and intent—the silent vocabulary horses speak every second. From that unique vantage point he developed a system that makes the invisible connections within a herd and between horse and human visible. He calls it **Herd Dynamics**. For me, it is a **compass**.

When Kerry and his wife Daphne landed at Schiphol in 2024 for their first Dutch clinic, I met two sincere and gentle people. Since then, Kerry and Daphne have become dear friends, and I am grateful for the collaboration we share. Kerry captivated the audience from the first to the last minute. During the practical analyses, riders and owners marveled at how he picked up the tiniest signals and translated them into clear, compassionate training choices. That moment showed me

HERD WIRED

that **Herd Wired** is not just another horse book; it is an invitation to look at connection differently.

That invitation now rests in your hands. In these pages you will meet the horse beyond muscle and conformation: the horse as sensory strategist, emotional mathematician, and partner whose greatest gift is authenticity. Kerry demonstrates that behavior we label as problematic often arises from a mismatch between our goals and the horse's survival logic. He also shows how to rewire the connection, not through force but through informed empathy.

My hope is that while reading you experience two things: first, the sigh of relief when something you have always sensed finally finds words; and second, the spark of responsibility that true insight brings. When we learn to see both the strong and fragile threads in our horses, we can weave training paths that lighten the load rather than add weight. That is the quiet revolution waiting in these chapters.

Open the door and step inside with soft eyes. The herd is already speaking; **Herd Wired** will teach you to listen.

Enjoy the read,

Nieske Pohlmann Equine Behavior Specialist & Trainer, New Forest pony breeder, event organizer Founder, **PAARDenLUISTEREN** www.paardenluisteren.nl

AUTHOR'S NOTE

As Alfred Hitchcock once said: "I have a feeling that inside you somewhere, there's somebody nobody knows about."

That's how I feel about every horse I meet.

I didn't write this book to change your mind, or endear you to my way of thinking, for I have no such expectations. My intention is to share with you that which I have learned so far, an invitation for your curiosity to look a little deeper by providing a lexicon of understanding for that which is felt but not always seen. Built not just on experience, but on timeless observation, emotional humility, and the kind of presence that comes from being vulnerable enough to learn.

I intend this to be a field companion for anyone standing at the threshold of understanding. And for those already deep on the trail, walking beside a horse whose truth they've chosen to hear.

Maybe, just maybe, it becomes a doorway for you, too—Not only into the mind of the horse...But into your own.

Thank you for walking this trail with me. My hope? That this work enriches your journey as it continues to shape mine, and within these pages you find your way to be the bridge, and not the block.

—Kerry

INTRODUCTION

What Herd Wired Really Is...

This is not a training manual. This is a raw, honest, and deeply personal journey into the unseen—into the interior world of the horse and the intricate emotional maps that shape how they think, feel, and connect.

Herd Wired: In Pursuit of Discovery is not a how-to. It is a field companion for the curious and the committed. A living document of observation, insight, and experience. It is shaped by years spent tracing horses through the wilderness of instinct and the structure of domestication—from untamed herds to high-performance arenas.

Along this trail, we are not just exploring answers to questions before they're asked. We are illuminating the natural patterns, threading the needle that knits the fabric of life, and bridging the gap between the natural herd dynamic and the domesticated world.

Each entry—whether it began as a field note, a case study, a blog reflection, or a quiet realization at dawn—has been curated not just for its information, but for its emotional resonance. These are signposts on a winding path of discovery. And like any true path, it does not always run in straight lines—but it leads somewhere meaningful.

This book is part memoir, part field journal, and part emotional codex. It invites you behind the curtain—into the developmental trenches, the defining moments, and the living thought process behind the evolution of Herd Dynamic Profiling™ and Sensory Soundness™.

These pages weren't born as a book. They began as scribbled margins, shorthand thoughts, and half-formed philosophies captured in countless notebooks before they slipped away. But over time, they began to form their own rhythm—an internal logic. A voice. A vision.

This is the voice of that work.

What you're holding is not a polished doctrine. It is not a methodology in bullet points or a tidy system of training steps. What you're holding is something living:

- A bridge between the science of nature and the art of horsemanship

- A lens through which to see the horse as both teacher and mirror

- A philosophy of feeling, and a practice of presence

This companion doesn't teach technique—it reveals the thought behind the technique. It traces how emotional patterns form, how herd-wired behaviors manifest, and how surface sameness can conceal entire emotional worlds beneath.

These pages are more than words. They are breath and memory, experience and emotion, Stitched together by time, reflection, and the deep current of lived connection.

This book is for those who believe the horse-human bond is more than mechanics. That it is communication through feel, clarity through emotional rhythm, and discovery through presence.

So, what is Herd Wired?

It's the notebook never meant to be shared...But the one that might matter most.

About the Images

All photographs in this book were taken by either Kerry or his artist wife, Daphne Thomas, during fieldwork, personal journeys, and the lived experiences that shaped this story. The pencil sketches are one-of-a-kind renderings drawn from Kerry's original research photographs—capturing horses in the wild and during private evaluations. Guided by Daphne's creative direction, these illustrations carry her artistic sensibility and emotional intuition, helping set the tone for the journey ahead.

These images are not signposts to the words beside them, nor mere decoration. They are trail overlook places to pause, lean on the fence rail of the page, and take in the country around you. Sometimes they show the horse. Sometimes they show the world that shapes the horse. Because the herd's story is never just the horse's—it is the wind in the grass, the shadow of the hawk, the movement of elk in the far valley, the quiet tension of predator and prey sharing the same horizon.

Placed throughout the book, they are meant to widen your view. To remind you that the study of herd behavior is also the study of life's greater pattern—how many lives move together, each influencing the other. In this way, the photographs and sketches become part of the language of the trail itself, speaking to the whole wild picture as Mother Nature intended.

Trailhead: Understanding Herd Structure

The Architecture of Belonging

There are moments in life when you step into a place you don't yet understand, and everything inside you quiets down. You don't know what it is you're looking at—but something tells you it's looking back.

For me, that moment happened in the high country of Montana, just west of the Crow Reservation. I had been tracking the Pryor Mountain Mustangs for days, watching from a respectable distance. On one quiet morning, perched just below a ridgeline, I watched a band of horses emerging from the trees in single file. Not hurried, not scattered. Moving as one.

It wasn't just motion I saw. It was message.

From where I sat, I couldn't tell who the leader was. That was the first clue.

I expected the "alpha" to stand out, to assert herself in obvious ways. But there was no hierarchy written in muscle or momentum. There was something else holding them together—*an emotional gravity*.

The horse at the front didn't look back to see if the others were following. She didn't need to. They *already were*.

This, I came to understand, was herd structure. Not as humans define it, but as horses *live* it.

Nature Doesn't Count Horses—She Connects Them

In our world, we like to organize. We assign numbers, build charts, and expect uniformity. But in the wild, a "herd" isn't defined by how many horses you see.

It's defined by the *emotional connections* between them.

Herd structure is not just a social ladder or a pecking order—it's a living, breathing *architecture of belonging*. It's not about dominance or obedience. It's about *fit*. Emotional fit.

Every horse holds a place in this structure not because of force, but because of *need*. Roles within the herd are shaped by emotional intelligence, sensory soundness, and the natural push and pull between strength and vulnerability.

When the puzzle pieces click into place, harmony is felt.

When they don't, the entire picture becomes blurry—horses fragment into pairs, stress rises, and a state of emotional limbo sets in.

This is as true in the wild as it is behind a fence.

From a Distance, We See Horses—

But Up Close, We Feel Their Roles

Let's pause here for a moment, because this is one of the most important things I can tell you:

You can have a group of horses without having a herd. You can have horses living together without living in structure. And you can have a "well-behaved" horse who feels deeply alone.

Just because a group shares space doesn't mean they share emotional cohesion.

True herd structure is composed of **distinct emotional roles**, fulfilled instinctively:

- The **Lead Mare**: not loud, but undeniable. She steers the herd through timing, subtlety, and strength of presence.

- The **Adjunct Horse**: the glue that binds. A high-level communicator who bridges the herd leaders with the body of the group.

- The **Mid-Level Horses**: adaptable and numerically dominant, making up ~50% of all horses in natural populations.

- The **Lower-Level Members**: the most emotionally vulnerable, reliant on peers to interpret the world for them.

- The **Lead Stallion**: a figure of protection, appearing intermittently, often from the periphery, more shadow than spotlight.

And then there are outliers—horses with such elite emotional intelligence and sensory soundness that they can function **independently of the herd**. These rare individuals express what I call **Independent Nature**.

Each of these roles is shaped by the internal balance of IHD (Individual Herd Dynamic) and GHD (Group Herd Dynamic), two psychological engines we'll explore more deeply as the trail unfolds.

But for now, just know this:

The herd doesn't ask horses to be perfect. It asks them to be *necessary*.

Emotional Camouflage: The Herd's Survival Code

In predator-prey ecosystems, visibility is vulnerability. But the horse doesn't just rely on physical camouflage. It relies on *emotional camouflage*—a seamless blending of roles and rhythms so that no single individual becomes a beacon for danger.

In small herds, this becomes especially critical. Emotional intelligence—not size, speed, or strength—is what keeps the group safe.

The horses I observed didn't scatter when a sound echoed in the valley. They moved *together*. Each horse responding to a flick of an ear, a shift in weight, a breath held or released.

Leadership was not declared. It was *felt*.

The Adjunct Horse might guide the young one forward, while simultaneously shielding the Lead Mare from standing out. And if the Lead Mare fell, the Adjunct would rise—not by ambition, but by nature.

This is what I mean when I say horses *fill the space they are needed in*. Roles can shift in response to stress, but only for a time. Long-term misalignment leads to collapse, often mistaken as behavioral issues.

But these aren't "problem horses." These are **unmet roles** searching for a structure to hold them.

Behind Every Behavior Is a Belonging

When a horse acts "out," we often ask, *What's wrong with him?* But rarely do we ask, *Where does he belong?*

Belonging isn't a luxury in the herd—it's a survival mechanism. And it doesn't come from obedience. It comes from emotional reciprocity.

In the wild, horses are ousted from the herd not because they're bad, but because they threaten the balance of harmony. In domestication, they don't have that option. They can't leave. So they internalize. They fracture. They fragment. Or they fall silent.

We can easily mistake that for calm.

But it may be something else entirely.

The View from the Hoof

In my early days of fieldwork, I had the great fortune of being mentored by a Crow elder. We walked together, we listened together, and I learned to let the land speak before trying to explain it. He told me once:

"Where your eyes may deceive you, your instincts will guide you."

That became my mantra for discovering the emotional world of the horse.

Feel what you see. See what you feel.

What I've learned since is this: when you take your **view from the hoof**—from *their* level, through *their* psyche—you begin to recognize a truth that changes everything:

A herd is not a number. It is a contract. A promise between emotional roles to protect, support, and adapt to one another in a living web of mutual dependency.

And when we—humans—ask to be part of that herd, we must first ask: *what role do I fill?*

What Comes Next

The trail ahead will guide you deeper into these ideas. We'll meet several unique herd dynamic characteristics and explore their varied—and often interwoven—roles along the way. We'll examine how psychology and environment influence behavior, and how we can work *with* the emotional operating system of the horse, rather than against it.

But before we begin, pause here. Take a breath. Look around.

You're not just entering a book. My purpose here is far deeper than that.

You're entering the world of the horse *as I have discovered it to be*—a place where research, philosophy, and theory converge into concept and practical application.

Here, language itself carries layered meaning. Things that *seem* the same often hold *very different truths*. And the more closely you listen, the more the subtle becomes visible.

This is the unseen world, revealed by feel.

Welcome to *Herd Wired*.

PART ONE

Laws of Nature: Field Lessons & Foundations

This section contains the foundational groundwork—original philosophies, raw field notes, and emotional explorations that shaped the concepts of Herd Dynamic Profiling™ and Sensory Soundness™.

These are the unfiltered ideas formed in the quiet moments of discovery, still bearing the fingerprints of their creation.

Here, we step into the earliest trailheads of this work—where instinct meets reflection, and raw observation becomes meaningful insight. These pieces are more than musings; they are **landmarks** on the developmental journey, forming the emotional and intellectual roots of everything that followed.

Let this opening part serve as a reminder: If we wish to explore the world of the horse, we must do so from the horse's view of the world.

Trail Marker One: The Emotional Operating System

What It Means to Be Herd Wired

At the very core of the horse's psyche is a wiring for mutual dependency—an invisible thread that binds the herd together through the strengths and the vulnerabilities of its members. Ironically, it is not individual strength that forms the true glue of the herd construct, but rather the communal strength found within individual vulnerability that connects horses to one another—and to us.

Before a horse is an athlete, a partner, or even a student, it is first and always a sentient being. A creature of feeling, not just motion—absorbing its world through instinct, associative memory, and the moment at hand. **That's where our journey begins.** Not with control. Not with performance. But with the unseen mechanism that governs all of it: *the operating system running the machine.* What I call *being herd wired.*

This isn't just a poetic turn of phrase. It's a framework—a lived truth in every horse, whether wild on the range or groomed for Olympic sport. *Herd wiring* is the silent architecture that shapes how a horse forms relationships, reads energy, interprets stimulus, and decides what to do next. Before a horse reacts, it processes. And before it moves, it interprets.

To understand horses—*truly* understand them—we have to start here. Not with what they do, but with how they feel their way through the world.

The Invisible Conversation

There's a gap between stimulus and response that is often mistaken for mere reflex. But the horse isn't reacting blindly. It's computing emotionally.

What we call "instinct" is often complex emotional interpretation occurring in real time. The horse reads the room before it responds to it. And within that space, between what is felt and what is done, lies everything that matters.

The conversation never stops. Between horse and horse. Horse and human. Horse and environment. It's not a conversation of commands—it's one of feeling. A quiet but profound language that speaks through breath, posture, and intention.

To connect with a horse, we must become fluent in this emotional language.

Sensory Soundness™: The Living Operating System

A horse can be structurally flawless and still fall apart under pressure. Why? Because emotional soundness must precede physical expression.

This is the heart of **Sensory Soundness™**—the horse's ability to interpret sensory input without being emotionally overburdened by it. It's not about desensitizing the horse to the world. It's about teaching them *how to read it*.

Sensory Soundness is dynamic. It rises and falls with the environment, herd structure, emotional leadership, and even the weather. A horse may seem sound in one setting and completely overwhelmed in another. That doesn't make them broken, it means their emotionally driven sensory processing is situational.

We cannot train consistency into a horse's reactions. We can only nurture clarity in their perception. And clarity is the bedrock of soundness.

Interpretive Ratio & Psychological Pace

Every horse has an *interpretive ratio*—a balance point between how much sensory data comes in and how much emotional capacity they have to process it. When that balance is tipped, the horse isn't disobedient—they're overwhelmed.

And that ties directly into a concept I call *psychological pace*—the speed at which a horse emotionally interprets its environment relative to how fast it's being asked to move. We must understand that ultimately the sensory system is charged with psychologically clearing space for the body to move through, like a blocker making a hole for the running back.

If we push the horse faster than their psyche can process, we outrun the emotional system. The body may be in motion, but the mind is left behind. The result? Miscommunication, stress behaviors, and a breakdown in trust. Any horse that "outruns" their ability to effectively clear space will experience mental fatigue at increased rates.

In high-performance horses, this synchronization between mental and physical pace is a marker of elite potential. It's not about athleticism, it's about how they handle their own sensory input *at competitive speed*. There is a difference between an athletic horse, and a horse that is an athlete.

Leading Through Emotion

The most valuable gift we can give the herd-wired horse is not control. It's *emotional clarity*.

You can't command understanding—but you can *become* the landmark in their emotional landscape. When your intention, presence, and energy align, you provide a reference point. A sense of stillness in a moving world.

The horse doesn't need you to take charge. They need you to become congruent—so they can feel safe enough to interpret freely.

This is what it means to lead through the emotional operating system. To become the fixed star in their moving sky.

The Six Sensory Zones

The horse perceives its environment through six sensory zones—each one tied to emotional awareness, social proximity, and survival logic.

- **Zone 1 – Forward**: Primary linear focus; driven by the *Independent Herd Dynamic (IHD)*.

- **Zones 6 & 2 – Left and Right Shoulder**: Side recognition; governing spatial sensitivity and confidence.

- **Zones 5 & 3 – Left and Right Rear**: Subconscious absorption; monitoring both herd and environmental energies.

- **Zone 4 – Directly Behind**: The "ghost zone"—where trust must replace visibility. Governed by intuitive memory.

Zones 2–6 are regulated by the *Group Herd Dynamic (GHD)*—the subconscious radar that reads the world behind, beside, and beyond what's visible.

When the zones are sound, the horse reads the world with fluency. When they're unsound, the horse reads chaos. And chaos—real or perceived—activates survival logic – fight or flight.

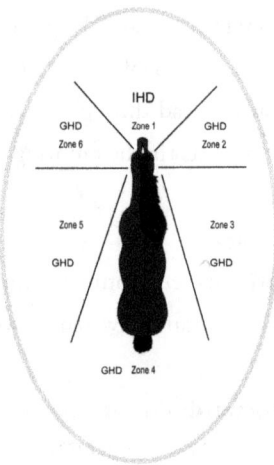

Sensory Overload and the Cost of Unsoundness

Sensory Soundness isn't about shielding the horse from experience. It's about strengthening their ability to interpret it.

When sensory input outweighs emotional filtering capacity, the horse enters a flooded state. We see this as:

- Spooking

- Shutting down

- Defiance

- "Bad behavior"

But what's really happening is stress overload; a nervous system crash— a bandwidth exceeded.

And that's why we must never confuse calmness with comprehension. Stillness is not always clarity. And motion is not always understanding. It is easy to misunderstand the emotional lexicon of expression, for we tend to see what we want to see.

Final Reflections

I remember watching a young colt once, standing at a gate while wind whipped across the pasture. His herd mates grazed calmly, but he was alert—*not afraid, just processing.* His ears, his eyes, even his breathing—all measured, reading the environment like a live wire.

He wasn't frozen.

He wasn't flinching.

He was computing.

That's the moment I understood—truly understood—what it means to be herd wired. There is a system running the horse. Not mechanical. Not programmable.

Emotional.

And when you begin to see that... you never stop seeing it.

Field Takeaways

- **Interpretation precedes reaction.** Physical expression is emotional exhaust.

- **Sensory Soundness™ is dynamic.** It fluctuates with environment, relationships, and internal processing.

- **The interpretive ratio matters.** Horses don't "blow up" without cause—they overload.

- **Psychological pace governs connection.** Don't outpace the horse's mind.

- **Clarity is leadership.** Emotional alignment builds interpretive trust.

- **You are the landmark.** Become a fixed point in their sensory map.

Quick Guide: Signs of Sensory Soundness vs. Unsoundness

Indicator	Sensory Soundness	Sensory Unsoundness
Attention	Soft, shifting between points of interest with ease	Fixed stare, hyperfocus, or scanning without settling
Response to Novelty	Investigates, adjusts posture, resumes rhythm	Overreacts, freezes, or avoids without recovery
Breathing	Even, deep, matches body rhythm	Shallow, rapid, or inconsistent with movement
Ear Movement	Independent rotation, tracks multiple stimuli	Locked forward or pinned, tracking one stimulus only
Body Flow	Movement matches mental state; transitions smooth	Movement outpaces or lags behind emotional readiness
Social Interaction	Engages fluidly with other horses	Avoids or clings, showing imbalance in herd role

Field Tip: Sensory Soundness isn't about being unreactive—it's about interpreting the world without losing emotional rhythm. Look for *balance* between awareness and relaxation, not the absence of reaction.

Trail Marker Two: Foundations of Thought

The Field Journal Mindset

Before it ever became a method or a model, my work began with something far less polished—moments of wondering. Not proof. Not protocol. Just a question scratched in the dirt: *What am I really seeing?*

The **Laws of Nature** files—those weather-worn pages from field journals and dusty case studies—were never intended for the public eye. They were mine. Scribbled insights. Fragments of intuition. Notes from the gut long before I had language to explain them. But in their rawness lives an honest record of how this work has grown: not from theory first, but from *experience first*.

If Trail Marker One taught us some basic understanding of the horse, this Trail Marker pulls back the curtain on how I learned to understand the horse. It's a glimpse behind the scenes. A walk into the wilderness of thought before it had names. **This is not instruction. This is inquiry.** Part science. Part soul. And all horse.

Seeing What Isn't Seen

"The moment you know what you're looking for is the moment you risk missing what's really there."— *Field Note, 2009*

True observation isn't about identifying behaviors. It's about resisting the urge to define too soon. It's about waiting. Watching. *Feeling the truth beneath the obvious.*

Early in my work, I started jotting little notes in the margins of my files—

- *"Appears calm, but energy tension sits in the ribcage."*
- *"Listening with the body, but not receiving with the mind."*
- *"Wants to connect, but doesn't know how to communicate it."*

These weren't training plans. They were *emotional footprints.* Traces of something deeper—things the body revealed long before action confirmed it. What began as feeling-based guesses eventually revealed consistent behavioral patterns. Not rooted in conformation or gait mechanics—but in *interpretation, association, anticipation,* and *trust.*

That's when the Laws of Nature stopped being scattered observations and started forming a map. Not a list of instructions—but a lens to look deeper. To wait longer. To *stop pretending I already knew.*

Philosophical Fragments from the Field

These are phrases pulled from the dirt. Truths that rose up not in books, but in breath shared with horses.

- *"Instinct is only a starting point. True intelligence is learned emotional interpretation."*
- *"A horse does not spook from what it sees—it spooks from what it doesn't understand about what it sees."*
- *"There is a difference between willingness—and the inability to say no."*

- *"Trust is not the absence of flight—it's the absence of internal conflict about whether to flee."*

- *"Environment is not a backdrop. It is a participant in the emotional story."*

The Law of Intention

Behind every healthy horse-human connection I've witnessed, one thing is always present—*and it's often overlooked.* That thing is **intention**.

Not pressure. Not command. But *presence with purpose.*

It's an unspoken invitation that says: *"I see you. I don't need to force understanding. I'm here until you want to meet me."*

This law is the heartbeat behind everything—*Herd Dynamics, Sensory Soundness™, trauma recovery, social bonding.* A horse may submit to pressure, but they *respond* to intention. Intention builds connection where pressure might only build compliance.

And I've seen it time and again: the most honest breakthroughs didn't come from clever exercises. They came when I stopped asking for anything at all. When the horse realized that I was simply there—*listening, without an agenda.* That I didn't need performance. Only presence.

That's when the wall breaks. That's when *vulnerability opens,* and with it, *true behavioral clarity.* Relationships are only as honest as the vulnerabilities that bind them.

The Wilderness Never Lies

In the wild, no one's keeping score.

There are no medals, no mirrors, no theories. There is only truth—written in movement. The twitch of an ear. The flick of a tail. The decision to flee or stay still. *Everything speaks.*

And the message is always the same: **Survival is emotional before it is physical.**

It's not just size or strength that keeps a horse alive. It's *how clearly they interpret the moment.* The wild horse doesn't waste emotional energy. They store it. Shape it. Deploy it with precision. Those who can't manage that economy of emotion? They put themselves—and the herd—at risk.

Reading the Emotional Map

Look closely, and you'll see choreography—not chaos. A foal doesn't just follow its dam—it learns rhythm. It studies emotional timing. A lead mare shifts her breath or flicks an ear, and the herd adjusts. Not out of obedience, but through survival instinct and learned life-skills. Because in the wild, measured response is survival. And survival depends on what we might call emotional reasoning: emotional intelligence through association—intellect without reason. I know you're not here, but I don't know you're gone.

Emotional Energy: The Real Currency

Emotional energy isn't abstract—it's measurable. It's a currency. Every horse carries an emotional "bank account." Every experience is a transaction.

- **Deposits**: clarity, calm, confidence

- **Withdrawals**: confusion, fear, miscommunication

Leadership makes deposits. Dominance only takes.

A horse that's constantly drained cannot process, trust, or respond with clarity. They're not being difficult. They're trying to survive an emotional deficit and battle through mental fatigue.

Linear vs. Radiating Emotion

Horses express emotional energy in two main ways:

1. Linear Expression (*Individual Herd Dynamic – IHD*)
- Directed focus

- Singular goal or target

- Confrontation, pursuit, direct problem-solving

- Task-specific intention

2. Radiating Expression (*Group Herd Dynamic – GHD*)
- Environmental absorption

- Monitoring multiple stimuli simultaneously

- Emotional awareness and cohesion

- Herd-orientating, communication hub, indirect problem solving

The herd dynamically sound horse shifts fluidly between these modes, adapting to pressure without losing rhythm. The more fluid the shift, the more confident the horse.

Herd Movement as Emotional Signature
Watch a wild herd move.
What you're really seeing isn't just locomotion—it's *shared emotional frequency*. When one horse changes breath, the group adjusts. Their movement is not just motion—it's murmuration in muscle and memory. Like a flock of birds carving shapes in the sky or a school of fish pulsing in unison, the herd moves as one through emotional synchrony.

This is not obedience. It is instinctive, sensory-based trust, a natural rhythm of survival communicated without words. Not from hierarchy. From connection.

Every member carries part of the herd's emotional weight, a key link in the hierarchical chain. Every movement is a message. And when you tune in, you stop seeing legs—You start seeing *language*.

Survival by Subconscious Design
In nature, hesitation can be fatal. Sensory Soundness™ becomes not just a strength—but a survival skill. A horse who interprets clearly is not just safer—it's *valuable*.

These are the true leaders in a herd. Not because they push others—But because they *steady them,* providing balance within a swarm of chaos.

Their emotional clarity becomes contagious, and the herd flows like water around them. The greatest desire of any horse is harmony with their environment, and contentment with their peers.

Domesticated Confusion

Now take that same emotionally fluent horse... and put him in a stall. Take away his herd. Mute his sensory channels. Fill the space with static, but no signal.

What once flowed now ricochets. Emotion with no channel becomes frustration, frustration becomes stress. We can call it "behavioral." But the root is *orientation loss*—emotional dislocation.

We didn't break the horse. We broke the map.

Emotional Energy in Our Hands

The wild may not be visible in the domesticated horse...But it is never gone.

That emotional operating system, the one refined over generations—is still there. Still running. Still interpreting. Still seeking clarity.

Our job is not to suppress it, but to *learn how to speak to it.*

We are not training motion. We are shaping meaning. We are not managers of behavior. We are interpreters of emotion.

Field Takeaways

- **Curiosity is connection.** Let go of your need to name everything.

- **Behavior is memory in motion.** The body may forget, but the horse never does.

- **You are always making a deposit—or a withdrawal.** Choose intention over pressure.

- **Processing speed is survival speed.** Emotional bandwidth is not a luxury—it's life.

- **The wild never leaves the horse.** Thus we must learn to meet them where it still lives.

Trail Marker Three: Mapping the Invisible

Sensory Zones, Emotional Space & the Physics of Pressure

The Shape of Space

Every horse has a body. That part is obvious. But just beyond the body—hovering like a magnetic field—is something unseen but deeply felt: **the sensory landscape.**

This landscape doesn't follow symmetry or measurement. It isn't fixed. It shifts. It breathes. And to the horse, it's everything.

If we want to understand behavior, we have to start here—not with the body, but with the *space around it*. Not with feet or hands, shoulder angle nor back-length, but with *feel*.

I didn't come to this realization through theory. I came to it through experience—field sketches, silent observation, horses whose reactions didn't make sense until I stopped thinking in distance and started thinking in *emotion*.

What I eventually mapped were **sensory zones**. But the map isn't drawn with numbers and measurements—it's drawn with feel and sensitivity. These zones expand and contract depending on emotional state, trauma history, relationship, and environment. One day, a horse's zone might be wide open—welcoming contact. The next, that same zone might be locked tight to the body, protected like a wound.

Assimilation is an inherent key for survival; a horse adapts to physical changes and assimilates to emotional ones. The use of equipment in training becomes learned behaviors through adaptation, the use of emotional intention teaches through assimilation, providing the horse with the life-skills to adapt. One is applied; the other is communicated.

Sensory Mapping isn't about control. It's about learning to *feel* emotional space as a form of communication.

Emotional Geography: A New Way to See

Imagine the horse as a topographical map. Each zone is an emotional region:

- **Zone 1**: The frontal channel—focused, targeted, where intent meets the world.

- **Zones 2–6**: The periphery—rear, sides, blind spots, subconscious reception, herd surveillance, and environmental absorption.

Now imagine that each zone is *malleable*, expanded by confidence, shrunk by fear. Trauma, herd dynamics, human handling, even environmental stressors... all of these influence how the zones behave.

- A horse that "spooks" easily? Might be experiencing sensory overload in Zones 3 and 5.

- A horse that "won't move off the leg"? Might be guarding a collapsed Zone 2.

- A horse that clings to you, or can't focus? Might have lost all periphery—a symptom of what I call **herd drift.**

When zones begin to collapse, a horse doesn't just lose balance. They lose *orientation*. They aren't resisting the world—they're retreating from it.

Herd Drift: When the Map Fades

Some horses are born with naturally balanced sensory systems, *sensory soundness*. Others aren't. But all horses can lose zone balance and sensory efficiency when they face emotional overload—especially if it's chronic or unresolved.

This drifting away from balanced sensory awareness is what I refer to as **herd drift**. It is closely related to but should not be misinterpreted as Herd Isolation Syndrome. Herd Drift is situational and sensory zone specific. – [Herd Isolation Syndrome is total horse, complete disconnection from the natural herd construct.]

- Clingy behavior

- Total withdrawal

- Hyper-reactivity

- Shut-down obedience

These are not disobedient traits. They're signs of a *zonal system under emotional pressure.*

The horse isn't being difficult. They're *disconnected*. They're losing touch with the emotional map that once made sense.

And without *leadership through emotional clarity*, the zones don't come back on their own. They don't expand again. They **calcify.** At that point, the horse hasn't just lost connection with you—they're struggling to orient with their world.

You Are the Landmark

The good news?

You don't need to "fix" the zones. You simply need to become a **landmark** in the sensory map—a steady point of reference. Not the controller. Not the driver. The *constant*.

The sensory system isn't built for submission—it's built for orientation. Horses don't want to be controlled. They want to *feel safe* enough to stretch toward you.

When your energy is consistent, congruent, and emotionally honest, you become that anchor point in their otherwise chaotic world.

Pressure doesn't restore balance. **Presence does.**

Cushion, Collision & Compression
The Physics of Emotion

Pressure isn't just physical. It's emotional. And the horse feels it long before it's ever applied.

Before your leg makes contact, before the gate opens, before your grip tightens—Your horse has already received the signal. And made a decision about how to interpret it.

That decision lives in the space between stimulus interpretation and response. That space is the **Cushion,** the psychological workspace.

Cushion: The Emotional Buffer Zone

Cushion isn't distance. It's *bandwidth*.

Two horses can be nose-to-nose. One is grounded. The other is on the brink. It's not how close they are, it's how clearly they're interpreting pressure.

Cushion is the emotional **breathing room** between input and response. It's what allows a horse to *choose* rather than *react*.

- The wider the cushion, the more time the horse has to process.

- The narrower the cushion, the more immediate the reaction.

Horses with high Sensory Soundness™ typically have generous cushion space—they *feel more*, but they *process more clearly*. Those with a contracted cushion are more likely to collect emotional stress, like air in a balloon. Eventually, that balloon will burst.

Compression: When Cushion Fails
Compression happens when cushion disappears.
It's not always dramatic. It often hides in plain sight:
- A horse that performs mechanically but mentally checks out

- A delayed response to cues

- A sudden explosion after weeks of good behavior

- A freeze—where the horse goes still, but not soft

Compression is bracing. It's the nervous system trying to cope with unprocessed input. It's the storm before the collapse.

Collision: The Breaking Point
At the end of compression is **collision**.
This doesn't always mean a buck or bolt. Sometimes it's emotional implosion. Sometimes it's shutdown.
- Collision is internalization of unprocessed stress, when pressure exceeds capacity.

- It's when reaction becomes reflexive, not chosen.

- It's the horse hitting the raw edge of emotional elasticity.

And here's the most important truth:
Most behavior issues are not about refusal. They're about cushion loss—emotional fatigue that's been misread as defiance.

Interpersonal Pressure: The Human Factor
Pressure doesn't only come from cues. It comes from *us*.

Our heart rate. Our thoughts. Our emotional weight. The horse feels all of it and symptoms include:
- A scattered mind
- A rush to perform
- A hidden frustration
- An unclear intention

These create **non-physical tension**. The emotionally sensitive horse will respond—not to our actions, but to our emotional rhythm.

The question isn't always "What did I do? "Sometimes it's "What did I bring with me into the moment?"

Herd Dynamics in Motion

In the wild, cushion is reinforced constantly through **Group Herd Dynamic** (GHD).

- The lead mare maintains emotional space, not just direction.
- The adjunct to her absorbs environmental pressure.
- Young horses learn from the ebb and flow how to negotiate emotional terrain.

It's a flowing system of emotional calibration. But in domestication, *we* become the herd. And too often, we bring collision instead of cushion.

How to Build Cushion

You can't drill cushion into a horse. You *grow it* through rhythm, timing, and emotional honesty.

The five essentials for building an elastic cushion are:

- Consistency without rigidity
- Pressure with permission to retreat
- Space to interpret, not just respond
- Time between cues
- Feel instead of force

Cushion is not something we demand. It's something we *nurture* and *protect*.

The Silent Collapse

Some of the most "well-behaved" horses are the most compressed emotionally. They comply. They respond. They look like "good horses." But inside—they are collapsing.

These horses haven't accepted pressure. They've surrendered to it. And that surrender isn't peace. It's emotional depletion.

Our job is not to produce obedience. Or incite coping mechanisms. It's to protect *interpretive freedom*.

Field Takeaways

- **Zones are emotional, not just spatial.** They reveal how the horse feels—not just where they are.

- **Cushion is emotional clarity.** The more space to interpret, the more safely to choose.

- **Compression is a silent crisis borne of emotional stress.** It builds slowly. It speaks softly. It breaks loudly.

- **Herd drift is sensory disorientation.** Not defiance—*disconnection*.

- **You are the landmark.** Be steady enough for their internal map to find you.

- **Motion is not the goal.** Shared *intention* is.

Quick Guide: Cushion, Compression & Collision

State	What It Is	How It Looks	What to Do
Cushion	Emotional breathing room between stimulus and response	Horse pauses to interpret, then responds with balance	Maintain steady presence, give time to process
Compression	Cushion narrows; horse collects unprocessed pressure	Mechanical responses, delayed reactions, rigid posture	Reduce stimuli, slow the pace, restore space
Collision	Cushion collapses; pressure exceeds capacity	Sudden blow-up, freeze before explosion, or emotional shutdown	Stop the ask, retreat to a calm state, rebuild slowly

- **Field Tip:** A "well-behaved" horse can still be compressed. Look beyond compliance—true cushion allows *choice*, not just control

Trail Marker Four: When Connection Breaks

Herd Isolation Syndrome,
Pressure Interpretation & the Journey Back to Belonging

The Loneliest Room is a Crowded One

There is no wound quite like emotional disconnection. And in the horse, it rarely arrives in a dramatic entrance. Instead, it creeps in quietly—like mist slipping beneath a barn door. A longing expression in the eye. A pause that lingers too long. A glance that goes *through* you, not *to* you.

It's not resistance. It's **social uncertainty**.

What I've come to call **Herd Isolation Syndrome (HIS)** is not a disease in the traditional sense. It's a psychological condition where the horse becomes emotionally removed from the herd structure—whether that herd is equine or human. It's not just a loss of connection. It's a *fracture of self-awareness*.

And the part that's often misunderstood:

A horse separated emotionally from the herd doesn't become more independent. They become **untethered**.

The Ghost in the Barn Aisle

HIS rarely announces itself during training. It shows up in the *in-between moments*.

- The way a horse enters a new space, rigid, hesitant but hollow.

- The stillness that feels less like peace and more like absence.

- The horse that stands beside you... but never really sees you.

Some become ghosts. Others become shadows. But in both cases, the light of connection has dimmed.

They may still perform, still respond to cues, still "do the job." But the performance is no longer a shared experience. It's a coping mechanism. They are not with us—they are *getting through us*.

And the cause? It's not always trauma. Sometimes it's chronic confusion. A constant mismatch between what they're being asked...what they're capable of and what they're emotionally prepared to give.

This isn't disobedience. It's **displacement**.

Not Alone, but Lonely

One of the cruelest ironies of Herd Isolation Syndrome is that it doesn't require physical solitude.

A horse can be **"in a herd"** and still be *alone*. We must be mindful that the number of horses do not comprise a herd, it is the emotional connection between them.

They may be in a herd of ten. In a bustling barn. Next to us in the stall. But emotionally, they are unreachable.

Because proximity is not presence. And partnership is not proximity.

Horses don't bond by standing next to each other. They bond by feeling *safe* inside shared space. And when that safety is broken—when attrition wears away clarity—connection dissolves.

Just like a person can feel invisible in a crowded room, a horse can feel **isolated in a full pasture**, in the barn, at a show, with us...

And unlike us, they can't walk out the door. So they walk inward. And carry the loneliness with them.

Rebuilding the Bridge

The good news? HIS is not a life sentence. It's a **call for repair**.

But the repair doesn't come from technique. It comes from *emotional rehabilitation*—a patient, steady reintroduction to shared rhythm, clarity, and trust.

What does that look like?

- **Intentional stillness.** Sitting quietly together with no task in mind.

- **Non-demanding interaction.** Mirrored breathing, grazing walks, free movement.

- **Relational anchors.** Pairing with emotionally sound horses or calm spaces.

- **Predictable emotional rhythm.** Letting them *feel* where you are, always.

And above all—**becoming worth returning to.** If a horse has disconnected from you, it is often because your presence no longer offered emotional harmony.

We must stop asking, "Why did they leave?" And start asking, "Did I ever truly invite them to stay?"

Processing Pressure: The Missing Conversation

Pressure is not the problem. **Interpretation is.**

Without pressure, nothing grows. But without *understanding*, pressure becomes a threat.

We've all been taught to think in terms of "How much pressure should I apply?" But the better question is:

"How is my horse *processing* the pressure I applied?"

Because the same cue can feel like guidance to one horse...and like a weapon to another.

Pressure Is Not Force—It's Information

To the emotionally stable horse, pressure is data.

- "Can you move here?"

- "Can you pause with me?"

- "Can you stay present through this moment?"

But to the overwhelmed horse, pressure is *confusion*. And confusion becomes fear. And fear becomes resistance.

We must stop using pressure like a hammer. And start offering it like a *conversation*.

The Five Types of Pressure

Pressure isn't always physical. And the horse's sensory system reads every form of it:

1. **Spatial** – Proximity, body position, and line of sight

2. **Rhythmic** – Repetition, timing, or pacing

3. **Emotional** – Internal energy and intention; emotional im-

pulse precedes physical action

4. **Task-based** – Requests for movement or mental engagement

5. **Environmental** – Wind, noise, smell, terrain, novelty

Each of these is processed through their unique sensory system—Filtered through instinct, experience, and static in the *emotional frequency*.

The Interpretive Equation
Every "ask" is an equation:
Orient + Identify filtered through Interpretation = Response
If interpretation is clear, the response is grounded, purposeful. If interpretation is disrupted, the response becomes reactive.

We don't shape the response by demanding it. We shape it by **supporting interpretation.**

The Myth of Desensitization
We must stop mistaking **numbness** for development.
There is a profound difference between:
A horse who has *learned to process* pressure.
And one who has *learned to ignore it.*
The latter may appear quiet...But inside, they are fraying.
Desensitization is not the goal. Discernment is.

Pronghorns- herd cross-study

Leadership That Listens

True leadership doesn't just apply gravity. It listens to the echoes in the room.

Ask yourself:

- Did the horse return to emotional baseline after the cue?

- Did the eyes soften or go vacant?

- Did the body stay fluid—or brace?

- Did I leave space for interpretation, provide an emotional escape, or did I chase obedience?

Leadership isn't about control. It's about **emotional fluency.**

The Pause Between Ask and Answer

So many moments are lost in the *rush* between cue and reaction. But the most powerful teaching lives in the **pause**.

Let the horse *feel* the question long enough to find their own answer or to guide them to it.

That's not softness. That's **coaching emotional intelligence**.

Shaping Belief, Not Just Behavior

Every time you apply pressure, you are shaping more than movement.

You're shaping:

- Whether that pressure is safe

- Whether your presence is trustworthy

- Whether mistakes are punished or supported

- Whether forward is a cue... or a consequence

You're not just influencing behavior.

You're writing belief into the nervous system.

Field Takeaways

- **HIS is emotional, not environmental.** Horses don't isolate because they're bad. They isolate because they're unseen.

- **Presence without participation is a red flag.** Look for the ghost or shadow in your horse's behavior.

- **Clinginess isn't connection.** Desperation and partnership feel different—even when they look the same.

- **Pressure is not the enemy.** Misinterpretation is.

- **Support interpretation.** Behavior is the visible part of a deeper emotional equation.

- **Discernment over desensitization.** You want a horse that feels —not one that fears or ignores the feeling.

- **The pause is the teacher.** Don't rush the answer—honor the space between; life skills matter and carry forward.

- **You are the landmark.** Be emotionally steady enough for them to return to, psychological consistency allows for assimilation in varying environments. They outsource to you for emotional anchoring.

Some horses don't "leave" the herd because they want to. They isolate when it doesn't feel to stay.

Let your presence be the invitation. Let your "ask" be a question worth answering. And let your connection be the anchor during emotional tsunamis.

Quick Guide: Five Types of Pressure & How They're Processed

Type	Description	Healthy Processing Looks Like	Overload Looks Like
Spatial	Proximity, body position, line of sight	Adjusts position calmly, reorients with ease	Avoids, crowds, or braces against space
Rhythmic	Repetition, timing, pacing	Anticipates pattern, stays soft in motion	Rushes, stiffens, or breaks rhythm
Emotional	Internal energy, intention, mood	Matches your tone, remains attentive	Mirrors tension, becomes unsettled
Task-Based	Requests for movement or focus	Attempts willingly, explores options	Resists, offers "autopilot" responses
Environmental	Wind, noise, scent, terrain, novelty	Scans, investigates, re-engages	Overreacts, freezes, loses attention

Field Tip: Pressure isn't the enemy—confusion is. Before adjusting *how much* pressure you use, adjust *how clearly* it's being interpreted.

Trail Marker Five: The Psychology of Herd Structure

Emotional Architecture, Nervous System Roles & Becoming the Portable Herd

The Nervous System in Motion

A herd is not just a collection of horses. It is an emotional **ecosystem**. A living, breathing, self-regulating *nervous system* composed of individual processors—each with their own rhythm, awareness, and emotional contribution.

From the outside, it may look like movement. But on the inside, it's *communication*.

Every twitch of an ear, shift of weight, or extended pause is part of an invisible choreography of interpretation. A thousand micro-decisions happening in the space between bodies.

The herd is not built on dominance. It's built on **discernment**.

Emotional Hierarchy, Not Rank

There are two primary psychological roles that shape the internal function of a herd:

1. **Independent Herd Dynamic (IHD)** This is the forward-driving presence—purposeful, singular in focus, confident in motion. The IHD is the emotional arrow that makes up roughly 75% of the male psyche.

2. **Group Herd Dynamic (GHD)** This is the emotionally expansive presence—sensitive, watchful, relational. The GHD is the emotional net and makes up roughly 75% of the female psyche.

These aren't ranks. They are *processing styles* shaped by the demands of herd survival and mutual dependency.

They are in essence the left and right sides of the emotional brain—one seeking focus, the other scanning for collective coherence. One leads the movement. The other holds the space in which that movement makes sense.

And together, they build the **architecture of sustainable survival** in uncertain environments.

Rethinking Leadership

We've been taught to look for the "alpha." The bold. The bossy. The one who moves others out of the way.

But the true lead horse is rarely the loudest. It is the one who holds **clarity under stress**.

Often a mare in wild herds, the lead horse isn't commanding. She's *transmitting calm.*

She leads not through force, but through:

- Emotional congruence

- Predictable rhythm

- Strategic Stillness

- The quiet authority of **being unfazed**

Leadership in a herd is less about control—and more about emotional *stability*.

Some horses carry the emotional load for the group, especially common in domestication. They are the early warning system, the interpreters, the sentinels.

- They feel the wind change first.

- They detect the energetic shift in another horse before it becomes behavior.

- They absorb the ambient tension others ignore.

These are the **emotional barometers**—the *first chair* in the orchestra.

In the wild, they share this responsibility with others. But in the domestic world, they are often left to shoulder it alone.

When removed from a functioning herd, they don't just lose companions. They lose **co-regulation**.

And it leaves them frayed, hyper-vigilant, emotionally over-tasked.

You Are the Herd Now

In captivity, natural herd dynamic doesn't disappear. It simply **transfers**.

That role—the regulating presence, the energetic rhythm, the organizing system—now belongs to **you**. Our very presence makes us part of the equation; we can be part of the at-large environment and viewed as potential predators or allow ourselves to be vulnerable enough to be absorbed. If you are pretentious you are viewed with skepticism.

Your horse always processes you through the lens of herd dynamics:

- Are you emotionally predictable?

- Do you offer calm or confusion?

- Do you hold your space, or collapse into theirs?

They don't care about your training method. They care about your **emotional coherence**.

If you feel like static, they will pull away or rush through. But if you feel like structure, they will align, harmonize, seek cadence.

The human becomes the herd. And with it, the responsibility of *regulation*.

The ultimate connection is when the horse's movement runs through your intentions.

When the System Breaks

Most "behavior problems" are not acts of rebellion. They are symptoms of **structural absence**.

- The horse doesn't know who is leading.
- Their emotional role isn't defined.
- The human sends conflicting signals.

And in herd life, confusion equals vulnerability. Static equals danger.

In the absence of structure, the horse will do one of three things:

1. Invent their own system
2. Submit in body but disconnect in mind
3. Become emotionally feral—reactive, guarded, checked-out

They don't do this out of disobedience. They do it to *survive*.

Restoring the Emotional Architecture

You don't need to dominate the herd. You need to **anchor** it.

Ways to become the regulating presence:

- Establish **predictable rhythm** in your presence and intention
- Create **emotional boundaries** without force
- *Listen* for the horse's **processing style**—do they scan or target?
- Allow room for **both IHD and GHD expressions**
- Read the **moment**, not just the method

And most importantly: Be the place where **neutral replaces static.**

Because for the horse, harmony and contentment aren't manifested through dominant control. They are realized through the **clarity** that manifests through purposeful navigation of uncertain environments.

Field Takeaways

- **A herd is a "nervous" system.** Every horse processes uniquely as part of the whole.

- **IHD and GHD are rhythms, not ranks.** Understanding your horse's processing style unlocks connection.

- **True leadership is emotional regulation.** The lead horse holds the line by staying grounded.

- **In domestication, the human becomes the herd.** Your emotional cadence shapes their behavioral response.

- **Structure is emotional security.** Not a system of rules—but a rhythm of trust.

Force is not a characteristic of leadership. Presence is. And when you are present, your horse won't just follow. They'll **join**—not because they have to… but because they feel safe enough to. How information is delivered affects how it is received, our emotional inflection forges the path that our words follow.

Trail Marker Six: Fragile Nature

Emotional Disruption, the Collapse of Self, and the Long Way Home

Breaking From Within

There are moments in a horse's life—sometimes subtle, sometimes seismic—when something inside begins to falter. A phenomenon, while uncommon in nature, is an unfortunately not-so-scarce passenger in domesticated life.

Not the body. The framework.

The invisible scaffolding that holds together identity, rhythm, trust, and emotional fluency starts to wobble. And sometimes, it collapses altogether.

It doesn't always happen with drama. Sometimes it happens in silence.

A blank stare. A missed cue followed by human urgency. A sudden unwillingness to move forward—not out of defiance, but as if some tether has quietly snapped.

I've come to call this **fragile nature**—the moment when physical development and emotional maturity fall out of sync.

A horse may look the part—growing, training, advancing—but inside, their emotional world lags. Sensory changes become sticky, expression heavy. They perform the movements, but not the meaning. They connect, but without enrichment. They follow the "leader," but not the relationship.

It's like the difference between an acquaintance and a friend. There is interaction… but not affection.

When the Mind Can't Keep Up

Many horses don't disconnect because they're unwilling. They disconnect because they're overwhelmed.

The horse whose body moves one way while the mind goes another is a flashing sign that internal discord is mounting. The space between the physical ask and the emotional answer fills with stress. With static. With a creeping sense of something missing.

Eventually, the horse starts running on autopilot—checking the boxes, but not checking in.

This is the slow erosion of emotional resilience. The byproduct of a herd-wired mind trying to navigate a domesticated world that doesn't always allow time to process—a frayed connection, like a flickering light in a thunderstorm.

Because emotional pressure without space to interpret can disrupt the sensory sequence itself—turning connection into confusion. And when emotional processing is skipped, when sensory sequences are jumped over, interpretation fractures. Conclusions are assumed. Rushed.

Pressure Without Pause

Every horse is born sensitive. It is not a flaw. It is their greatest strength.

Their sensitivity is a living antenna system—receiving sound, space, motion, intention, ambiguity, and contradiction all at once.

But no system can receive forever without time to sort.

When input becomes relentless—when the pressure never stops and the horse is asked to move without time to breathe—something inside begins to jam.

At first, the horse tries to compensate. Then, they try to assimilate. Then, they begin to shut down.

Not because they're stubborn. But because their operating system is too congested to continue.

They don't need less work. They need more time to **be**.

Cracks in the Frame

You can often spot a fragile internal state before it becomes a full break—if you learn to read the emotional language.

Watch for:

• Disjointed reactions – abrupt energy spikes that don't fit the moment

• Emotional heaviness – slow interpretation, as if the mind is buffering

• Mismatched hesitation – stillness not from choice, but confusion

• Loss of herd sync – withdrawal from group rhythms and muted body language

• Erratic fear – explosive responses to minor or invisible inputs

These aren't signs of a "difficult" horse. They are the markers of a horse who's lost their internal map.

Trying to navigate a world they struggle to anticipate, there is a disassociation even with the familiar.

Rebuilding the Emotional Framework

Healing a fragile nature isn't about restoring what was. It's about creating something new.

These horses need more than training. They need **emotional repair**.

And that begins with you:
- Be the first thing that makes sense again
- Be the steady rhythm in a chaotic world
- Be the one who shows up—not just physically, but emotionally

You don't need to do anything dramatic. Just be consistent. Present. Predictable. Honest.

Invite the horse to rediscover themselves in your presence—without rush or hurry, without judgment, without agenda.

The result should never outweigh the process.

And when they begin to feel safe enough to offer something new—even a glance, even a breath—you'll know the new framework is starting to hold.

The Beauty of the Reassembled

Some of the most emotionally intelligent horses I've known were once deeply fractured.

An unknown past paired with misunderstood reaction cannot be soothed without patient enrichment. Most horses in domestication struggle to find a natural fit within herd construct and spend their lives developing coping skills. They camouflage it well, keeping insecurity hidden from predatory eyes—instinctively, eyes like ours.

But there is yet a way forward...

Humans who encourage them with **practical life skills**—not tricks, not desensitization, not just training routines—equip them with the tools they need to begin to assimilate properly.

We want the horse to manage their sensitivity with purposeful response—not be emotionally dormant, numb, or conditioned. **Sensitivity is the gateway to healing.**

To nurture this in training, it is essential to provide them a mental escape route—a space to run to, not a wall to crash into.

These horses can learn to interpret collateral pressure. Not to ignore it, but to **discern** it. To find purpose in movement, not just reaction.

Because when movement flows through emotional intention—mind and body become one.

That's the goal. Not obedience. But **synergy**.

Field Takeaways (Part I)
- Fracture is emotional overload, not laziness or resistance
- Lag and confusion are early warning signs—don't dismiss or overcorrect hesitation
- Consistency matters more than technique—you are the repair space
- Rebuilding doesn't restore the old—you're helping them build anew
- Fractured doesn't mean finished—these horses often find their way again

Bison - herd behavior study

The Moment Everything Changes... or Doesn't

There are times, often many, where things can abruptly shift along an otherwise calm journey.
- A new environment
- A traumatic event
- A botched interaction
- Or even... a tiny mistake in timing

And in that moment, the internal pattern either holds—or it doesn't. Was it an interruption? Or a fracture?

Interruption: A Detour with a Return

An interruption is a break in rhythm, not identity. The horse stumbles emotionally—but finds their footing again.
- They hesitate, recalibrate, then return to trust
- They question, but don't stop asking

- They remain tethered, even if tentatively

This is recoverable. It's a bruise, not a break.

Fracture: The Collapse of Trust

A fracture is different. It doesn't pause the system, it splinters it.
- The horse no longer sees the world the same way
- They stop reaching out and start guarding their own inner space
- The emotional currency between you becomes emotional debt

And here's the hardest part: It's not just a shift in behavior patterns. t's a shift in the way patterns are **learned**.

The fractured horse doesn't just resist training. They resist any **association** with it—often expressing vehemently within environmental proximity of it.

This makes assimilation impossible and adaptation possible only through strong measures.

Gaps in the sensory sequence create a bumpy road—the horse seems like they have to continuously **reengage** with any task, ask, or target. The sensory system switches into defensive mode, *who is attacking me and from where?*

Recognizing Fracture

If you see:
- Sudden regression after strong progress
- Over-compliance without engagement
- Freeze responses before explosions
- Vacant expressions despite stillness
- Reactions that seem disproportionate

You're not dealing with disobedience. You're touching an unhealed scar.

How Fractures Form

Fractures emerge when:
- Pressure exceeds interpretive capacity
- The horse is asked to override their internal signal
- They're left alone inside emotional distress
- Confusion is punished, not redirected
- A moment of trauma becomes a lifetime of anticipated defense

And from that point on, part of the horse never leaves the moment of rupture.

Preventing the Break

You can't prevent all trauma. But you can prevent many fractures.
- Don't chase compliance over clarity
- Don't force forward when the horse is caught in confusion
- Don't stack asks without room for reset
- Pause. Always pause. The **pause** is where safety lives.

The Way Back

To reconnect with an emotionally splintered horse:
- Restore choice before demanding task
- Allow expression before refining movement
- Establish harmony and contentment before applying pressure
- Offer a new lens—one built not on obedience, but upon shared leadership

Fractures are not flaws. There is strength in vulnerability.

Go back to a place that's just before they hesitate. Ask for nothing. Just share space with them.

Don't be the one who only calls when you need something. Redirect the anticipation to a place where **response replaces reaction**.

The mutually dependent nature of the horse relies heavily upon **outsourcing**—searching for the response to a question they cannot answer themselves is a natural binder, and it is an inherent tendency of the heart-wired mentality.

Outsourcing serves as a **bridge** and helps maintain harmony and contentment in the natural world. When we isolate the horse from their natural herd construct, we isolate them psychologically, exposing both strength and vulnerability.

When we do that, we do not remove their natural tendencies. We become **accountable** to them.

Be present. Be steady. Be grounded in emotion. Intentional in presence, sincere in expression.

Field Takeaways (Part II)
- Interruption is a disruption. Fracture is a break in sensory sequence.
- Fractured horses protect themselves. Self-preservation is a default instinct.
- Healing requires emotional fluency, not control.
- Trust is found in shared leadership—not just compliance.
- The greatest gift is time

Quick Guide: Interruption vs. Fracture

State	What It Means	Signs in the Field	Response
Interruption	Temporary break in rhythm; identity intact	Hesitates but re-engages, stays mentally tethered	Pause, allow recalibration, resume slowly
Fracture	Break in sensory sequence; trust compromised	Vacant eyes, guarded posture, over-compliance, or explosive defense	Remove task, restore choice, rebuild emotional framework over time

Field Tip:

An interruption is a bruise—recoverable with time. A fracture is a break—requiring patient reconstruction of trust before performance.

Trail Marker Seven: Environmental Design

How Space Shapes the Horse's Emotional World

The Architecture of Emotion

Before a horse learns to perform, they must learn how to **process**.

Before they respond to us, they respond to **where they are**.

The environment isn't just a backdrop. It's an **active participant** in the emotional development of the horse.

Every horse is a product of emotional environment—not just what happens *to* them, but what surrounds them.

From stall to trailer, arena to pasture, the physical world becomes a **psychological map**.

And the design of that world...**matters**.

The Unseen Pressure of Space

Stalls. Round pens. Aisles. Trailers. Gateways. Corrals.

To the human eye, they're functional. To the horse's mind, they're **sensory experiences**.

Each space has its own emotional **signature**:

- Blind spots and echo points

- Angles of approach and escape

- Rhythms of light, airflow, and footfall

- Scent trails and memory imprints

What seems safe to us may feel like confinement. What feels neutral to us may feel noisy or unclear.

Because horses don't just occupy space—they **interpret** it.

They don't memorize floorplans. They create **emotional maps** based on sensory data and lived experience.

Designing for Sensory Success

Supporting **Sensory Soundness™** isn't about training harder—it's about *designing smarter*.

Ask yourself:

- Can the horse see where they're going?

- Is there a way to retreat or reset?

- Are exits clear—emotionally and physically?

- Is sensory input coming from one direction—or from all sides at once?

- Are stimuli predictable… or chaotic?

True confidence is built through **navigable space** and **sensory rhythm**.

Design isn't just aesthetic.

It's **psychological architecture**.

Routines vs. Ruts

Routine can be comforting—but when overused, it becomes *numbing*.

- Same space.

- Same task.

- Same sequence.

Over time, this dulls interpretation rather than deepening it. Repetition without **variation** leads to stagnation, not confidence.

To develop emotional agility, horses need safe opportunities to explore **variability**:

- Shifting the flow will curate fluency

- Introduce emotional *textures*; different inflections upon similar tasks

- Offer novel but manageable patterns; don't *always* turn right

A **stimulated** horse is a **thinking** horse. Stimulus should invite interpretation, not overwhelm it.

The Horse's Design System

The horse doesn't store tangible environments like we do. They store **emotional topography**, that which is associated with the tangible.

- The trailer where stress occurred is encoded.

- The gate that gave them a shock becomes a hotspot.

- The arena where they found balance becomes a sanctuary.

Their **psychological *ecosystem* system** keeps score.

They carry space with them—not in memory, but in **association**.

And those associations will influence every future interaction.

You are Part of the Environment

Perhaps the most overlooked part of the horse's environment...is **you**.

Your energy. Your timing. Your breath. Your rhythm.

You are not separate from the stall, the round pen, the trailer. You are *woven into it*.

The horse doesn't learn in isolation. They learn in **relational design**.

Which means:

- Your presence can create **cushion**... or **compression**.

- Your silence can be calming... or confusing.

- Your consistency can build trust... or trigger anxiety.

The horse isn't just reading space. They're reading **you**—as part of the space itself.

Designing With the Mind in Mind

To support emotional processing, we must design not for control—but for **interpretation**.

- **Flow matters more than finish**

- **Interpretation matters more than obedience**

- **Consistency matters more than symmetry**

Properly considered the environment becomes a *co-teacher*, allowing the horse to become *emotionally invested* in *learning* process.

Let the world around them teach stillness, clarity, rhythm. Let it speak before you do. If you need a fence rail to hold the horse in your space, you are not connected, you are asking for adaption to the physical environment, not nurturing assimilation to the emotional one.

Field Takeaways (Part I)

- **Environment is emotional.** Every space has a sensory fingerprint.

- **Good design supports interpretation.** Bad design demands obedience.

- **You are part of the environment.** Presence is architecture.

- **Routine without rhythm becomes restrictive.** Variation is learning's ally.

- **Design with empathy, not just efficiency.** Space teaches long before you do.

Wearing the World
Emotional Camouflage in the Horse

Present, But Not Seen

I once met a gelding at a rehabilitation facility. Beautiful. Healthy. Mannered.

He loaded, led, stood still. Soft eye, quiet body.

But something was missing. Not in his behavior. Not in his health. In his **eyes**.

He was present... but not **there**.

It was like looking through a window that had forgotten it was glass.

What I had encountered was **emotional camouflage**.

And once you've seen it, you can never unsee it.

What is Emotional Camouflage?

Emotional camouflage is the horse's ability to **blend** emotionally into their surroundings—especially in environments where they feel uncertain or overwhelmed.

In the wild, it's protective. It hides herd leaders from predators. It keeps the emotional current of the group from drawing too much attention.

But in **domestication**, it becomes something else entirely.

A horse appears:

- Obedient

- Calm

- Manageable

But inside, they've made themselves **emotionally small**. Unnoticeable. Even to themselves in that their sense of self-awareness shrinks.

This isn't shutdown. Shutdown is a collapse.

This is more subtle. More **sophisticated**. It is *performance without self-investment*.

The horse is moving, learning, responding—but not expressing.

They're playing the part, not living the role.

Why It Matters

Because when we mistake **camouflage** for **connection**, we build on top of an illusion.

Training that ignores emotional camouflage often leads to:

- Short-lived success

- Confusing regressions

- Sudden outbursts that "come from nowhere"

You hear it all the time:

- "She was perfect at home."

- "He's never done that before."

- "It just happened out of nowhere."

But it didn't come from nowhere. It came from **within**. From a place unseen, unheard, unaddressed.

How it Develops

Camouflage emerges when:

- Emotional input outweighs processing time

- Desensitization becomes numbing instead of learning

- The horse adapts to herd or human pressure without expression

- Vulnerability results in our ambiguity and their discomfort

Many foals learn this young—to mirror energy rather than offer their own. To adapt instead of interpret; it's not teaching, it's processing…

They stop reaching out, and start playing it safe, lest they be reprimanded for the imposition. Ironically, these early stages of expression will be counted upon as they mature for athleticism in competition. Squelched early, having to adapt to the strict timelines of domestication, competitive nature can be mitigated long before we ever "train it".

Field Notes: Signs of Camouflage

- A horse that "does everything right" but feels distant

- Eyes that are *empty* despite good health

- A disconnect between stimulus and response – automatic replies

- Sudden outbursts under new pressure

- A tendency to "go with the flow" until the flow breaks

These horses are often labeled:
- "Easy"

- "Quiet"

- "Mature beyond their years"

Is this a controlled calm? Or are they emotionally **invisible**?

Building Through the Camouflage

To reach the horse beneath the surface, you must **lead from presence**, not performance.

Here's how:

- **Go slow, then slower.** These horses aren't slow learners. They're slow *trusters*.

- **Stop rewarding obedience alone.** Look for curiosity, allow for it.

- **Watch the eyes, not just the ears.** The ears tell you attention. The eyes tell you truth.

- **Offer safe ambiguity.** Create environments where they can *explore* emotion without needing to *perform* it. If you're not working with the natural herd dynamic, you're working against it.

- **Narrate your presence.** Let them feel your intent before they feel your ask.

Their quietness might be the loudest thing they're saying.
Don't miss it.

Reflection

I didn't "fix" that gelding I met long ago. That wasn't the point. But I *did* witness something extraordinary.

One day, standing beside him in the round pen—doing nothing—he turned and looked at me. Not with his body. With his **eyes**.

And in that moment, I met the real horse.

Emotional camouflage doesn't just hide the horse from us. It hides the horse from *themselves*.

When we help them remember who they are—we've already begun the healing.

Because in the end...**Who** must always supersede **what**.

Quick Guide: Designing for Sensory Success

Element	Supportive Design	Stressful Design
Visibility	Clear lines of sight, gradual approach to blind spots	Sudden blind corners, blocked views
Exits	Multiple or clearly visible escape routes	Narrow, obstructed, or hidden exits
Stimulus Flow	Predictable, coming from one or two directions	Chaotic, coming from all sides at once
Acoustics	Soft, consistent background noise	Echoes, sharp or unpredictable sounds
Lighting	Even, natural or diffused light	Harsh glare, deep shadows, rapid changes
Space Shape	Allows turning, repositioning, and retreat	Tight, restrictive, or dead-end areas

Field Tip: Horses "map" spaces emotionally, not architecturally. Make sure the emotional map is one they want to carry with them.

"Reflect on where you've been, consider where you are, envision where you're going..." *kmt*

PART TWO

Living the Work: Essays, Reflections & Applied Insight

These are not chapters. They are waypoints—Markers along a journey that was never meant to be linear. Let's walk it together—one entry at a time.

In this second half of Herd Wired, we leave behind the foundation stones and step into moving water—into the living terrain where theory and practice converge.

These are the reflections, case notes, insights, and raw experiences gathered along the trail of a life spent chasing emotional rhythm in action.

Some were born in field notebooks. Some in the silence of wild herds. Others emerged in the breath between heartbeats—during a private session or a quiet walk.

Here, you won't find prescriptions. You'll find presence.

This is where the philosophy of Herd Dynamic Profiling™ becomes felt, where Sensory Soundness™ is no longer a definition, but a lived, breathing condition.

Each entry is a window. Some open outward. Some inward. All of them point back to the same truth:

That the emotional world of the horse is real, rich, and reachable—if we slow down long enough to listen.

Trail Marker Eight: Processing Pressure

Field Notes on Emotional Saturation & the Value of Retreat

The Fire and the Flame

There's a difference between **experiencing** pressure and **processing** it.

Just as there's a difference between **touching** fire...and understanding **heat**.

In the horse's world, pressure isn't just something applied—It's something absorbed. Interpreted. Emotionally transposed.

Most horses don't react to pressure itself. They react to the *anticipation* of it, a safety net webbed of emotional energy distribution to keep pressures from becoming entrenched points of emotional stress.

And when the mind begins to brace for something it doesn't yet understand, what we see is not response—but **preemptive defense**, the Anticipatory Response Mechanism.

This is **emotional saturation**: A state where the psyche is no longer learning...it's enduring.

At that point, it's not the pressure that matters. t's what the pressure **represents**.

To an emotionally healthy horse, pressure can become a **conversation**. To an oversaturated or fractured mind, pressure becomes a **threat**.

The Cushion Before Collapse

Processing pressure begins with **permission**.

Not the permission to perform—but the permission to **pause**.

The horse must feel safe to step back emotionally, not just physically.

This is where the **cushion** comes in.

Not just a theory, but a **felt experience**—a buffer zone between stimulus and response.

Some horses develop this naturally. Others must be taught—gently, patiently—that it's safe to:

- Breathe

- Hesitate

- Interpret

- Re-enter

Too often, we reward **compliance** over **comprehension**.

But comprehension requires time. And time requires patience.

Holding Space Within the Herd Dynamic

In the wild, common stresses are processed **collectively**.

A young horse might flinch—then glance to the herd, watch their response, and recalibrate.

The herd acts as both **"mirror"** and **"filter"**,... a *bridge*.

In domestication, **we** become the filter.

And if we are emotionally scattered, impatient, or unclear, we don't resolve pressure. We **amplify** it as stress.

Your emotional integrity becomes part of the horse's nervous system. You're not just applying pressure—you're modeling how to *exist* inside of it before common tensions become uncommon stresses.

In doing so, you offer the horse an **emotional vocabulary** they didn't know they were missing.

Retreat as Intelligence

Stepping away from pressure is not weakness. It is often the horses, and our, **wisest** choice.

A moment of retreat is not an evasion—it's a **strategy**.

We must not punish these moments. We must **honor** them.

Because when we do, we teach the horse that processing and controlled response is more important than obedience and reaction.

Processing stress isn't about "pushing through." It's about knowing **when** and **how** to step back.

Field Takeaways

To process pressure is to **feel**, **interpret**, and **learn**—not just to **survive**. Commonly experienced stresses feel like internal pressure that we can still move through, navigate, even use to our advantage. It must breathe, like letting steam from a teapot, lest the pressure put too much stress on the pot and it blows in dramatic fashion until the *pressure* is gone.

As you work with your horse:
- Look for signs of **emotional saturation**: glassy eyes, delayed responses, overreactions to small stimuli, or total emotional shutdown.

- Offer opportunities to **retreat** without consequence. Even a few seconds outside the pressure zone can restore clarity. Think *psychological intervention.*

- Understand: The goal isn't **desensitization**—it's **comprehension**.

- Let the horse processing time to understand what's being asked—not just **endure** it.

Processing pressure well is not about being **tough**. It's about being **available**—on *both* sides of the lead rope.

Context, Communication & the Clarity of Intention
The Horse Isn't Reading What You're Doing—They're Reading What You Mean

When the Unspoken Speaks
There's a moment every horseman knows:
You breathe…and the horse moves with you.
You think…and the horse prepares to follow.
That's not obedience. That's **interpretation**.
And it only happens when your **intention** is clear.
Horses don't follow cues. They follow **meaning**.
They don't look for commands—they look for **context**.
They are not interpreting your aids—they are interpreting **you**.

The Brain of the Horse Isn't Linear—It's Layered
To the horse, life isn't made of step-by-step instructions. It's a sensory **flow** of rhythm, emotion, energy, space, and pattern.

- Your posture has **tone**

- Your breath has **meaning**

- Your silence has **shape**

Everything matters.

And the more emotionally *present* you are, the less physically *loud* you need to be. Emotional communication is the lexicon of the horse.

The most effective cues are often the **quietest** ones—Because they carry the **clearest** intent.

Context Creates the Conversation

Ask yourself:

- Are you giving a cue… or inviting a **thought**?

- Are you managing motion… or guiding **emotional energy**?

- Are you speaking into a **void**… or relating to an **awareness**?

Without context, communication becomes **static**.

We repeat. We correct. We escalate. All in pursuit of a response we never earned emotionally, with thoughtful intention.

You can't demand connection when you haven't built a language.

Clarity Begins with Emotional Honesty

One of the hardest things for humans to do is be emotionally honest in front of others, but the horse, knows…

The horse already reads:

- If you're nervous

- If you're pretending

- If your confidence is borrowed
- If your kindness is conditional

They don't respond to how you want to be **perceived**. They respond to how you actually **feel**.

And that's where communication truly begins—not in your hands, not in your legs, but in your **intention**.

The Power of Unspoken Language

I remember a mare once—quiet as a shadow.

She didn't spook. Didn't resist. Didn't seem to care about anything.

People called her "easy."

But I felt she was just... **absent**.

She moved through the world like a ghost—her body here, her mind far away.

One day, I stopped *doing*. I met her in **silence**.

And for the first time, her ears flicked toward me.

Not because I asked—but because I stopped asking.

We didn't achieve anything "impressive" that day.

But we had a **conversation** perhaps she'd been waiting years to have.

Horses are fluent in silence. We're the ones who talk too much.

Intentional Communication Isn't a Skill—It's a State

Horsemanship isn't about cue-mastery. It's about **presence**.

Can you be clear, not because you remembered a technique—but because your **intention is clean**?

Can you lead with energy that **invites**, not insists?

Can you make space for your horse to say something **back**?

Because until you can hear them, you're not really speaking—you're just **directing**.

The Rhythm of Emotional Dialogue
Real communication is a **rhythm**.
- Give and take.
- Pause and pulse.
- Echo and response.

And sometimes, the most powerful thing you can do...
is nothing at all.

Field Takeaways
- Horses respond to **clarity of intention**, not complexity of cues.
- **Presence precedes technique.** They interpret who you *are*, before what you *do*.
- **Context is conversation.** Without it, training becomes static.
- The **quietest signals** carry the deepest meanings.
- **Emotional honesty** builds trust faster than skill.
- You don't train connection.

You *create space* for it to occur.

Quick Guide: Signs of Emotional Saturation

Sign	What It Indicates	Suggested Response
Glassy Eyes	Mind disengaged, emotional processing stalled	Pause, reduce demands, allow quiet reset
Delayed Response	Processing lag, cushion narrowing	Slow pacing, give longer pause between cues
Overreaction to Minor Stimulus	Emotional bandwidth exceeded	Retreat to familiar task or calm space
Total Shutdown	Collapse of interpretive sequence	Stop the ask entirely, focus on re-establishing trust
Preemptive Defense	Anticipates pressure before it arrives	Lower emotional tone, soften presence, reframe ask

Field Tip: Saturation isn't about weakness—it's about capacity. Restoring space to think is more valuable than pushing through.

Trail Marker Nine: Beyond Body Language

Listening Between the Lines

The Space Before the Signal

We often say that horses "speak" through body language. A flicked ear. A swished tail. A wide eye.

These are the signs we're taught to read.

But the deepest conversations aren't always seen—They're **felt**.

What we call body language is often just emotional residue—the echo of something the horse has already processed.

To truly communicate, we must feel the moment **before** the movement.

That pause. That stillness. That breath when energy shifts—but hasn't yet taken form.

That's where the real conversation lives.

Emotion Is the Root—Not the Result

Body language isn't the start of communication. It's the **byproduct**.

It is the shape that emotional energy takes once it passes through memory, perception, and experience.

If we train ourselves only to respond to what we *see*, we will always be a step behind.

But if we allow ourselves to feel what's beneath the surface—to attune to the **emotional temperature**—we begin to speak the language horses use in real time.

Between the Ears: A Different Kind of Listening

"Between the Ears" isn't just a vantage point. It is itself a discipline.

It's the art of reading a horse from the **inside out**—tracing the shape of thought, the nuance of intent, the moment when *choice* is forming.

You don't need to see tension to feel it. You don't need to wait for a tail swish to know something's off.

You already know. You've not forgotten how to trust that *knowing*, even though it makes us feel vulnerable. But that's a point, for once again we see therein lies the connective strength of vulnerability.

Field Companion Insight

The horse isn't hiding. They're speaking in a language we've unlearned.

Here are reminders to help you tune in:

- **Feel first. Interpret second.** Before you analyze what you're seeing, pause to sense what you're feeling.

- **Check for congruency.** A relaxed body paired with anxious energy is not truly relaxed. Learn to feel the differences especially when they are subtle.

- **Slow the moment down.** Connection often happens in the breath between cues.

- **Don't just observe your horse. Be with them.** That's where the dialogue begins.

Invisible Influence & The Power of Being
You Don't Always Have to Do Something to Change Something

The Unspoken Shift

There is a kind of presence that changes the room without saying a word.

A kind of energy that **invites** attention—not by demanding it, but by being deeply **settled** within itself.

Horses recognize this immediately. They feel it before they analyze it.

They're not reading your **actions**. They're reading how you **exist** in their space.

And more often than not, **that** is what makes the difference.

Invisible Doesn't Mean Passive

To influence a horse, you don't have to be active. You have to be **available**.

- Available to feel before you move
- Available to receive before you guide
- Available to stand still without becoming absent

This is **invisible influence**.

- Presence without pressure

- Invitation without insistence
- Energy without ego

Not all influence is visible. Some of the deepest shifts happen in the space **between** moments.

The Horse Doesn't Miss a Thing

In the wild, I've seen horses pick up the subtlest of shifts:

- The **intention** of a predator long before it moves
- The **internal tension** of a herd mate shifting in the wind
- The **breath** of the wind carrying a different story than the day before

Horses live in the realm of **nuance**. They survive by reading what's not being said.

So when we enter their space full of plans and agendas—they often brace before we even reach for them.

They're not resisting us. They're **absorbing** us.

Field Note: The Unseen Conversation

I once stood beside a stallion who wouldn't let anyone touch his head.

People had tried everything—ropes, food, trickery, time.

But still, he flinched when hands reached toward his ears.

I did nothing.

I didn't speak. I didn't reach. I didn't *ask*.

I just stood beside him, offering nothing but the honesty of my presence.

Breathing. Feeling. Needing nothing from him.

And after a long while, he lowered his head—just enough to rest it lightly against my chest as if to remind me he was there.

It wasn't a breakthrough. It was a **whisper**.

That wasn't about technique. That was about **being**. The smallest of things can bring the greatest of change, breaking an association even when we have no idea what that may be, often comes down to sharing space without it being a precursor to request.

Presence Is the Most Underestimated Skill in Horsemanship

We are a doing species. We measure progress in tasks, actions, checklists.

But horses don't.

- They measure leadership in the **tone** of your energy.

- They measure harmony in the **rhythm** of your presence.

- They measure contentment in the **space** you're willing to leave unfilled.

Because presence isn't just a **tool**. It's an **offering**.

You Are Always Communicating

Even when you're doing nothing—your horse is reading **everything**.

- Your rhythm

- Your emotional charge

- Your intention

- Your availability

- Your authenticity

They aren't waiting to be told what to do. They're listening for **what it feels like** to be near you.

And that…is where all connection begins.

Field Takeaways

- Influence isn't always visible. **Stillness can be louder than movement.**

- Presence isn't passive—it's **potent**. You don't have to act to have impact.

- Horses absorb **energy** before behavior. Your tone matters more than your tools.

- You are always communicating. Even your **silence** has shape.

- The horse doesn't respond to what you're doing. They respond to what it **feels like to be with you**.

- Being with a horse isn't about control.

It's about **relationship**.

Quick Guide: Micro-Signals vs. Macro-Signals

Signal Type	What It Is	Examples	Why It Matters
Micro-Signals	Subtle, early indicators of emotional change	Softening/hardening of the eye, brief ear flick, shallow breath, slight weight shift	Reveal emotional state before it becomes physical behavior
Macro-Signals	Large, visible expressions of emotional state	Head toss, tail swish, pawing, moving away, sudden stillness	Confirm a state that micro-signals have already hinted at
Integration	Reading both in sequence	Micro: ear flick → Macro: step away	Allows early intervention and prevents escalation

Field Tip: Macro-signals tell you what's happening *now*—micro-signals tell you what's *about to happen*. Train your eye for the small changes, and you'll shape outcomes before they escalate.

Trail Marker Ten: The Emotional Map

Navigation Without a Compass

The Map Beneath the Movement

Every horse carries a map inside them.

Not necessarily of the world around them—but of how they **feel** about it.

This map is not drawn with lines but by the brush of association. It's drawn with experience, painted with feeling.

With memory. With emotion. Within the emotional landscape, and what came next.

A corner of the arena is never just a corner. It might be:

- A pressure zone

- A moment of freedom

- A place where nothing ever happened—so it holds no emotional weight

The emotional map isn't about space. It's about **meaning**.

Landmarks of Emotion

Emotional mapping is a kind of internal geography.

The horse doesn't mark "places" by physical location—but by **emotional association**.

- A grooming stall isn't good or bad. It's a **memory center**.

- A trailer isn't stressful or neutral. It's an **emotional landmark**, shaped by previous experience.

If you want to change how a horse feels about a place, you must change the **emotional coordinates**. And that means layering new associations, breath by breath.

This is the essence of emotional cartography. You aren't leading a body through space—you're guiding a mind through memory.

Mother Nature's Laboratory

The Human as Mapmaker

You are always on the horse's map. The question is—**what kind of landmark are you?**

- Every interaction becomes a reference point.

- Every moment of pressure becomes a signpost: *Approach with caution... Safe to explore... Retreat now...*

And just like with a trail, the more **consistent** the terrain, the more confidently it can be navigated.

Horses don't ask for perfection. They ask for **predictability**.

They want emotional terrain that is legible. Familiar. Safe enough to explore.

Field Companion Insight

Here are ways to guide your horse through emotional terrain with intention:

- **Acknowledge emotional history.** If a horse hesitates, ask what their map is saying before you insist on progress.

- **Build emotional bridges.** Don't just get them through something. Walk with them through it. Let them redraw the territory with you.

- **Trust hesitation.** Pausing isn't disobedience. It's data. A moment where the map is fuzzy—and the horse is asking for clarity.

- **Be a predictable landmark.** Show up with emotional consistency. That's how trust is mapped and memorized.

Emotional Habituation & The Myth of Desensitization
What We Call Desensitization Is Often Emotional Numbness

The Danger of "Calm"
In much of what's labeled "natural horsemanship," desensitization is often seen as the goal. At least it seems that way to me.

We wave tarps. Slap ropes. Toss flags. And when the horse stops reacting—we call it success.

But is stillness always peace? Or have we simply taught them not to feel?

Sometimes stillness is a sign of shutdown. Not safety.

Owing to being herd-wired, there's a fine line in the equine psyche between **association** and **disassociation**. And we don't always notice when it's been crossed.

The Equine Psyche Doesn't Just Learn—It Copes
When a horse is repeatedly exposed to a controlled stimulus without time to process, they don't become braver.

They become **habituated** to it.

That means:
- They no longer interpret the stimulus
- They stop expressing curiosity
- They suppress emotional signals tied to it
- They risk dissociating from related environments

This isn't progress. It's **protective withdrawal**.

And the cost is often invisible—until it isn't.

Stillness Isn't Always Safe

We often reward the horse who "does nothing. "But what kind of nothing is it?

- Grounded stillness is peace
.• Frozen stillness is survival.

A horse standing statue-still—wide eyes, craned neck, breath held—isn't calm.

They're stuck in **Stage One of the Sensory Sequence: Survey**—a freeze response, trying to Orient.

Dishrag loose and perceptive...or rigid and stressed?

If we don't learn to read the difference, we disconnect from the very system we're trying to engage.

Emotional Habituation ≠ Emotional Healing

Desensitization, as commonly practiced, is often:

- Flooding the nervous system
- Creating tolerance through exposure
- Conditioning through repetition

But horses don't just learn behaviorally. They feel emotionally.

And when you suppress enough signals—you don't erase them. You **bury** them.

And buried signals don't disappear. They can **detonate**—later, and often without warning.

True enrichment comes from teaching life skills—tools the horse can apply in new and unpredictable situations.

It's the difference between surviving chaos and navigating it.

The value in desensitizing—*conditioning*—is often human-centric entertainment, engineered in peculiar, controlled environments.

But I, for one, care more about say, how a horse handles **herd isolation under competitive stress** than how they handle a tarp on their back.

Field Note: The Horse Who Couldn't Feel
There was a gelding I once evaluated.
His owner said,
"He's bombproof. He never spooks."
And he didn't. Not at tarps, flags, loud noises, or sudden movement.
But what I saw wasn't bravery. It was **emotional silence**.

- His eyes were dull

- His movement mechanical

- His curiosity gone

They hadn't trained the fear out of him. They had trained the **aliveness** out of him.

Over time, we didn't re-teach behavior. We re-opened **communication**.

He needed permission to feel again. Not pressure to perform.

What Horses Really Need Is Emotional Literacy
True development isn't about how much a horse can **tolerate**. It's about how much they can **understand**; life skills matter.

- Interpret stimulus, not just endure it

- Regulate emotion, not suppress it

- Express fear, not just hide it

This isn't desensitization. It's **emotional education**.

And it's the difference between a numb horse...and a connected one.

You Don't Want a Horse That Doesn't Feel

You want a horse who can:

- **Feel—and stay regulated**

- **Process—and remain curious**

- **Express emotion—and return to center**

That doesn't come from flooding the system. It comes from **partnership**. It comes from allowing sensitivity to discriminate between sensory inputs so the horse can continue to learn and not lump associative responses together.

And it begins when we stop asking:

"Why aren't you moving?" And start asking: "What are you feeling?"

Field Takeaways

- Desensitization often suppresses emotion instead of cultivating growth

- Stillness can be a sign of **freeze**, not peace. Learn to feel the difference

- Emotional habituation is not emotionally healthy nor is indifference growth

- Interpretation leads to resilience. Unnatural persistence can lead to shutdown

- You want horses to **feel**. Feeling is how trust is formed, allowing them to outsource

- Flooding conditions compliance—but is it entertainment at the cost of curiosity

- **Bravery isn't forced. It's fostered—forged through harmony and contentment**

- **Quick Guide: Reading the Emotional Map in the Field**

- **Field Tip:** An emotional map is dynamic. Read it often—small shifts in markers can predict big changes in behavior.

·Quick Guide: Reading the Emotional Map in the Field

Map Marker	What It Reveals	Field Check
Gaze Direction	Primary point of attention; potential trigger location	Is the horse looking *through* you, *past* you, or *to* you?
Ear Rhythm	Shifts in sensory focus between zones	Do ears move fluidly or lock in one direction?
Breathing Pattern	Synchronization of mind and body	Is breath deep and matching movement, or shallow and erratic?
Body Alignment	Comfort with environment and handler	Does the horse align to engage, or angle away to protect space?
Movement Flow	Cohesion between mental state and physical steps	Are transitions smooth or mismatched with emotional tone?

TRAIL MARKER ELEVEN: INTERRUPTION VS. FRACTURE

Moments That Change the Map

The Line That Breaks
Every horse carries a timeline.
And somewhere along that line, there are **interruptions**......and there are **fractures**.
At first glance, they might look the same:
- A sudden pause in development

- A change in behavior

- A disconnection in rhythm

But one is a **momentary stumble**. The other—a **permanent scar**.
- **Interruption** is a detour.

- **Fracture** is a disconnection.

And knowing the difference is the key to emotional recovery.

The Weight of a Moment

Interruption happens when something disrupts the horse's sense of emotional continuity:

- A sudden change in environment
- An unexpected injury
- A stressful haul
- The loss of a companion

There is confusion—but not collapse. The horse pauses... recalibrates... and often rebounds.

The emotional GPS still works. It just needs time to reroute.

But **fracture** is different.

Fracture occurs when the disruption isn't just external—it's **internalized**.

- The emotional system no longer trusts itself.
- The continuity of safety is snapped.
- The horse is not just confused. They're **disoriented from within**.

Living with the Break

Fracture isn't always loud.

In fact, it often looks like **compliance**.

The horse that goes along, that moves when asked, that stands still without spark—

That horse may not be trained. That horse may be **shut down**.

There's no curiosity. No questions. No emotional presence.

At some point, the horse **stopped asking**—because the answers were too painful.

Where an interrupted horse still seeks connection, a fractured horse often avoids it. Or worse—can no longer recognize it when offered.

Repairing the Invisible

Not all fractures are permanent. But healing them takes more than training.

It takes **reconnection**.

This work is quiet. It's slow. It's sacred.

You are not fixing behavior. You are **healing storylines**.

It begins with:

- Emotional consistency

- Safe, pressure-free space

- Recognition of the path behind the behavior

To help a fractured horse feel again, you must become a witness, not a mechanic.

Field Companion Insight

Not every pause is a break in the soul. But when the soul *is* broken, only presence—not pressure—can begin to mend it.

As you walk beside your horse:

- **Distinguish fear from confusion.** Fear withdraws. Confusion still reaches.

- **Don't rush reconnection.** Sit in the silence. Let them feel your presence before you ask for performance.

- **Create question-safe environments.** Fractured horses

may ask again and again: *"Are you safe today?"* Emotional "escape routes" are vital.

When you honor the difference between **interruption and fracture**, you're no longer training a horse. You're walking beside a survivor.

Rebuilding Curiosity
The Psychological Signpost of Safety

Curiosity Is a Feeling—Not a Task

Curiosity isn't just a behavior. In essence it's a **nervous system state**.

When a horse feels safe, they explore:

- They sniff

- They engage

- They follow sound instead of flinching from it

- They reach forward with presence, not just movement

Curiosity is a **signal**. It tells you: *The horse is here, now.*
And that means the system is no longer protecting—it's inviting.

You Can't Train Curiosity—Only Invite It

You cannot force a horse to explore. You can only create conditions where exploration is **welcome**.

That means:

- **Removing pressure**

- **Holding emotional space**

- **Giving permission to wonder**

The moment a horse asks a question, You've entered a new layer of trust.

That question is a gift. Don't rush past it. Creating purposeful, sensory stimulating and engaging obstacle courses, for example, is a tremendous way to thoughtfully curate an environment of life skill learning.

Field Note: A Look, A Step, A Pause

There was a filly I worked with once.

She didn't run. Didn't resist. Didn't react.

She had learned that stillness was safer than sensation. Her world had gone silent—internally and externally.

For three sessions, I asked her for nothing. I simply sat nearby. Breathing. Listening. Being.

And then one day—she looked.

Just a flick of the ear. A turn of the eye. A pause that held possibility.

She saw me. She wondered.

And in that spark of curiosity, she chose to feel again.

Curiosity Isn't Taught. It's Allowed.

In horsemanship, it's easy to chase *desensitization*. To train the *fear* out of the horse. To *push* through the response.

But true development comes from the **opposite**.

- Let them feel.

- Let them wonder.

- Let them pause.

Curiosity is what returns when fear no longer drives the system.

Rebuilding Curiosity Is Rebuilding Trust

When a horse becomes curious, they're saying:

- "I feel safe enough to explore."

- "I trust that I won't be punished for wondering."

- "I'm willing to step into the unknown—because I believe I'll be okay."

This is not compliance. This is confidence.
It is not control. It is **connection**.

Let Curiosity Lead

If your horse starts exploring something during a session—**let them**.

- If they sniff the cone—pause the pattern.

- If they watch the wind—don't pull them back.

- If they look at you—look back. Answer the question behind their eyes.

That's not distraction. That's emotional presence showing up in real time.

You're not losing focus. You're **gaining relationship**. If you don't give your horse time to find harmony in their environment, they will not find true contentment with their peers... namely you. You can only truly "train" from a space of harmony and contentment if you want that training to stick.

Field Takeaways

- Curiosity is within the psyche, a "nervous system" indicator of emotional safety.

- You can't train curiosity—you can only invite it.

- Stillness isn't always connection. But curiosity often is.

- Exploration isn't loss of control—it's evidence of trust.

- Every moment of curiosity is a moment of learning and can become a moment of healing.

- Growth begins when fear fades—and wonder returns.

- Follow and coach, train through their curiosity, not compliance. That's where the relationship lives.

Trail Marker Twelve: Behavioral Landscape

The Silent Sculptor of Behavior

Not all pressure is physical. Sometimes it's architectural.

A horse's behavior is not only shaped by how they are handled—but by the emotional architecture of the world around them.

Environment is never just background—it actively participates in your horse's experience.

From alleyways and layout flow to fencing, light angles, wind exposure, and water access—these elements may seem like logistics to us. To the horse, they're part of a living, breathing matrix that sculpts how they feel.

The Lens of the Environment

Imagine living in a world where every shadow matters. Where airflow carries information. Where sound isn't just heard—it's interpreted as either safety or threat.

This is the world your horse occupies. And in that world, environment is feedback.

When a horse walks into a space, they are scanning for:

- Escape routes

- Sensory clearance and posture

- Herd access or isolation, peer review

- Echoes from past experiences, associations

- Pressure points (seen or unseen)

If the horse doesn't feel emotionally safe in a space, you will never train them out of that feeling. You must design them into it.

Designing for the Mind, Not Just the Body

Too often, environments are designed for tradition or convenience—not for cognitive wellness.

Consider:

- A narrow alleyway may feel like a funnel of pressure to a spatially sensitive horse.

- A round pen beside heavy foot traffic may splinter emotional rhythm instead of focusing it.

- A stall with poor lighting may amplify confusion in a horse that processes visually.

Bottom line, design isn't decor. Design is direction for the sensory system.

The question is no longer *"Will the horse adapt or assimilate?"* It's *"What design will help this horse feel seen, safe, and supported?"*

Field Companion Insight

Environment doesn't just influence behavior. It pre-conditions and essentially prepares it.

Here are a few field principles to guide sensory soundness through spatial design:

- **Design with the individual in mind.** What calms one horse may overstimulate another. There is no universal blueprint.

- **Use environmental pressure points as a training tool—never a crutch.** A quiet space can foster early development, but overuse dulls the processing system.

- **Create zones of regulation.** Every horse needs a place to return to themselves—away from stimulus, demand, or interruption. A space to breathe built into training and competition is vital. *Let the steam out of the teapot before it whistles.*

The greatest designs are not the flashiest. They are often the quietest. The ones that allow the sensory system to exhale within the framework.

Mental Fatigue and the Recovery of the Mind
Not All Exhaustion Is Physical

Some horses don't stop because they're tired in their body—they stop because they're tired in their mind.

They're not worn out. They're worn down.

- Too much pressure

- Too many decisions

- Too little space to process

- Not enough emotional rest

These horses aren't lazy. They're spent. When we also consider that most horses will experience the sense of herd isolation, especially under high stress conditions, it's easy to see how mental fatigue sets in. I am willing to wager that you, just as I, have, and do experience *mental fatigue*.

Fatigue Can Look Like Disobedience
We misread the signs. We think:
- "He's not trying."
- "She's ignoring me."
- "He's being resistant."

But underneath the surface, the horse is saying:
- "I don't have the energy to manage all of this."
- "I've reached the end of my emotional *bandwidth*."
- "I'm shutting down to protect myself."

And when we respond to that shutdown with more demand—we don't wake them up. We drive them further away.

The Emotional System Has Limits
Just like muscles fatigue with use, the sensory-emotional system fatigues with overload.
- Too many tasks in a short span
- Too little recovery time between efforts
- Constant exposure without decompression

The result? Horses who go dull. Who stop offering. Who seem checked out or disengaged.

But what they really are...is emotionally overloaded. And when mental fatigue sets in before physical fatigue thresholds, the fine line that divides training from trauma has been crossed.

Field Note: The Horse Who Gave Too Much

I once evaluated a mare who had been called "quiet," "easy," and "willing" for years. She never resisted. She always tried.

But she was losing weight, grinding her teeth, and had developed a chronic pattern of ulcers.

When I met her, she didn't say "no." She didn't say anything. She just blinked. Stood still. And waited.

It wasn't obedience. It was exhaustion.

We didn't "fix" her with better training. We gave her back time. Time to recover. Time to process. Time to remember she was allowed to feel.

That's when she came alive again.

Rest Isn't Optional. It's Foundational.

We don't just need to train smarter. We need to recover more intentionally.

Recovery doesn't just mean turnout. It means:

- Emotional decompression

- Time to integrate experiences

- Freedom from constant interpretation

- Room to *be* without the demand to *do*

The psyche doesn't build new pathways under pressure. It builds them during rest.

Emotional Recovery Is a Relationship Practice

A mentally fatigued horse isn't resisting you. They're inviting you to slow down and feel with them.

When you:

- Pause before asking again

- Step back when they check out

- Allow a full breath cycle between transitions

- Shorten the session before the mind begins to splinter

You're not losing training time—you're gaining long-term capacity.

Field Takeaways

- Design influences behavior—before handling ever begins.

- Environment isn't passive. It shapes emotional rhythm.

- Emotional fatigue is not defiance. It's depletion.

- Overexposure can create mental shutdown long before physical fatigue sets in.

- Recovery is where growth takes root.

- Emotional regulation must be designed into the daily experience.

- Slowing down isn't going backwards. It's giving the mind permission to move forward.

Trail Marker Thirteen: The Gift of Going Slowly

Time Is Not a Training Tool—It's a Relationship Space

A Rhythm Beyond Speed

There is a pace at which horses live that has nothing to do with movement. It's not fast or slow in the way we measure things. It's a **rhythm of presence**. A **tempo of trust**. A way of being that says:

Let's be here now—together—before we go anywhere at all.

To the horse, going slowly isn't about doing less. It's about **being more aware**.

When Fast Comes From Us

We are creatures of ambition.

- We want results

- We want steps

- We want progress

But the horse doesn't think in objectives. They feel in **moments**. In the quiet interior of the equine mind, **rushing feels like risk**. To move forward before connection has formed is to **abandon the emotional anchor**.

Ask yourself:
- Have I earned enough presence for forward motion?
- Have I allowed enough pause for processing?
- Have I created enough space for this moment to mean something?

Because to the horse, *what you skip becomes what they fear.*

Field Note: The Day I Did Nothing Right

There was a wise old gelding—a seasoned soul who had seen both kindness and cruelty.

I came in too fast.
- Too much energy
- Too much intention
- Too much *me*

He didn't spook. He didn't fight. He just turned slightly away—a half-pivot of quiet dismissal. Still present in body...but **gone**.

No defiance. Just absence.

So I stopped. I stepped back. I sat on the rail fence and let the wind do the talking.

For twenty minutes, I did nothing. And then—he looked over his shoulder. One step. Then another.

Not because I asked. But because I allowed.

That moment reminded me: A step taken in trust is worth more than a mile made in haste.

Slowness Is the Language of Safety

To a horse, going slow is not laziness. It's **attunement**.
When you slow down, you're saying:

- *I'm listening*

- *I won't rush your process*

- *This moment matters*

- *You're not a task—I see you as a partner*

Slowness is how horses test the ground of trust. Not by how fast we go, but by how long we're willing to stay with them—before asking them to go anywhere at all.

The Tempo of Trust

Every horse has a unique **emotional tempo**—a psychological rhythm where harmony lives.

Too fast... and they retreat. Too slow... and they lose interest. But just right...and something **sacred** happens.

The emotions synchronize. The rhythm softens. Presence becomes mutual. Movement becomes a shared idea—not a command.

That's not just forward motion. That's **forward together**.

Field Takeaways

Here are a few truths about time and the horse:

- **Slowness is not a lack of progress.** It's the presence of **respect**.

- **Going slowly builds relationships, not just obedience.**

- **Rushing skips emotional steps.** What's skipped is often where insecurity lives.

- **Interpretation takes time.** Let the horse translate experience before they act on it.

- **Let the horse set the rhythm.** They feel when it's ready.

- **One aligned step in trust is worth more than ten disconnected ones.**

To walk with a horse is to enter their time zone. Not to pull them forward, but to **travel at the speed of understanding**.

That's where the real relationship is built—not in how fast we get somewhere...but in who we become together along the way.

Quick Guide: Pacing for Emotional Assimilation

Pace Element	Assimilation-Friendly Approach	Risk of Rushing
Task Introduction	One new variable at a time	Layering multiple changes at once
Cue Timing	Clear ask, pause for processing	Rapid succession of cues with no pause
Environment Change	Gradual exposure, controlled sensory input	Sudden, high-stimulus shifts
Progression	Builds on previous comfort zones	Skips foundational steps
Session Length	Ends on a calm, confident note	Ends in fatigue or heightened tension

Field Tip: Assimilation isn't about how slowly you *go*—it's about how fully the horse can *own* each step before taking the next.

Trail Marker Fourteen: The Psychology of Place

The Horse Doesn't Just React—They Relate to Place

Spaces That Carry Memory

Horses don't move through space the way we do. They don't simply occupy a location—they emotionally **map** it.

- They register thresholds.

- They record past experience.

- They assign safety or stress to terrain.

- They feel the weight of what happened *where*.

To the horse, every corner, gate, and fence line carries **emotional residue**. They're not just places. They are **signatures of meaning**.

Where It Happens *Matters*

The same cue might work in one corner—but fail in another. Why?

Because that corner once held fear. Because that gate led to separation. Because tension and stress still echoes in the threshold.

Place becomes memory. And memory becomes behavior.

If we ignore place, we misread reaction.

Field Note: The Invisible Trigger

A mare I once worked with would freeze—always in the southeast corner of the arena. No bolt. No flinch. Just... nothing.

The trainer called it defiance. The vet found nothing wrong.

But one day I stood there, quietly, alone—and I felt it.

That space carried expectation. Repetition. *Stress*.

She wasn't afraid of the **corner**. She was afraid of the **meaning** it had collected.

So we rewrote it. We rested there. We breathed there. We let new stories settle into old ground.

And slowly, it changed.

We didn't alter her behavior. We changed the **place it belonged to** through the associative aspect, *new skin on an old scar*.

Read the Emotional Map

If your horse:
- Tenses near the gate

- Rushes a corner

- Balks at the mounting block

- Stalls at the same fence line

Don't just train through it. **Trace the terrain** and work towards identifying what is associated with it. The expressions you see are the results of preceding emotion. We cannot remedy by treating with expression, we need to swim upstream.

Ask:
- What emotion lives in this space?

- What memory is attached to this ground?

- How can we soften the story here?

The breakthrough may not be in the *ask*, but in *where* it's asked.

Every Ride Writes the Landscape

Each interaction contributes to emotional geography. So let that contribution be gentle.

- End sessions in calm places.

- Revisit tough spots with softness.

- Use your horse's chosen rest zones as relationship zones.

- Let the landscape become a place they *want* to be.

You're not just guiding your horse through a space—
You're helping them feel safe inside their **emotional map**.

Field Companion Insight

- Horses emotionally imprint on physical locations.

- Place isn't neutral—it shapes meaning and memory.

- Re-patterning space can be as important as retraining behavior.

- Trust grows when we reshape not just what we do... but *where* we do it.

- We're not just riding *in* space—we're shaping the emotional story **of** it.

Echo Points — Side Trail

The psychology of place doesn't stand alone—it's woven through many of the emotional patterns explored along this trail. For deeper connection to these ideas, see also:

- **Trail Marker Six: Fragile Nature** – on how interruption and fracture can shape emotional associations with space.

- **Trail Marker Seven: Environmental Design** – on crafting environments that invite safe interpretation rather than demand compliance.

- **Trail Marker Fifteen: Psychology of Motion** – on how movement patterns through space reinforce or disrupt the emotional map.

Each of these entries connects back to the same truth:
Where it happens matters just as much as what happens—and sometimes, even more.

Trail Marker Fifteen: Displacement Behaviors & Silent Signals

The Horse Who "Seems Fine" May Be Saying the Most

Not all signs of struggle look like struggle.
Sometimes, they look like... nothing at all.

- A yawn after a stressful moment

- A shake of the head before trying something new

- A lick and chew that doesn't quite settle

- A sudden pawing, tail swish, or glance away

- A shift in focus to anything other than you

- A subtle tension in the entire countenance

These could be quirks. They could be random. They can also be *camouflaged displacement expressions*—the body's quiet cry for release when internal stresses gnaw like a lingering ache.

The loud signs of distress are easier to catch. But the soft ones? They're speaking, too. And often... they're saying the most.

When Emotion Can't Be Expressed, It's Displaced

Displacement happens when a horse experiences internal conflict—a clash between what they feel and what they're allowed to express.

They may:

- Want to move, but feel frozen

- Want to connect, but feel unsure

- Want to leave, but feel trapped

- Want to comply, but feel overwhelmed

With no clear outlet, the horse improvises. They redirect. They displace.

"I can't do what I feel—so I'll do something else instead."

This isn't defiance. It's negotiation. And too often, we miss the conversation beneath the calm.

Field Note: The Quiet Horse in the Round Pen

There was a gelding I once observed—fluid, responsive, textbook-perfect.

But between every pause and even transition, he pawed at the ground. Not once or twice—*every* time.

Some said it meant he was "releasing." Others called it boredom. My suspicion was something deeper: cognitive dissonance.

He wasn't relaxed. He was confused—trying to process triggers that didn't yet make sense.

In these moments, my default is to stop asking and start listening. I stood still. I closed my eyes, settled my breathing, letting whatever steam might be building in the kettle leak away before it whistled.

In that pause, the pawing slowed. His eye softened. His head lowered. And without a cue, he offered the next movement—calm, unforced, true.

He didn't need *more* training. He needed a moment to feel safe enough to be real.

We can't always pinpoint the cause. But when we build-in time-stops for the horse's emotions and senses to catch up, we help their mental cushion regain elasticity.

Reading the Displaced Emotion

If your horse:

- Licks and chews excessively after every cue

- Paws in moments of hesitation

- Rushes through tasks without grounding

- Swings their head without spooking

- Looks alert yet oddly absent

Ask yourself:

- Is this expression—or avoidance?

- Connection—or coping strategy?

- Responsiveness—or emotional redirection?

Sometimes, what looks like willingness is actually an emotional detour. And when we push through it, we teach the horse to hide what they feel instead of trust us with it.

Create Space for Silent Signals to Speak

To support a horse who displaces:

- Slow your pace and soften your presence

- Reduce simultaneous stimuli

- Notice the signal before correcting—seek to understand

- Let the horse orient, pause, and breathe with you

- Invite interaction rather than insist on performance

Displacement behaviors don't need to be "trained out." They need to be understood. They fade naturally when the horse feels safe enough to *just be*.

Remember: you can "touch" your horse without ever laying a hand on them. Entering their emotional space thoughtfully is as important as physical contact—sometimes more so.

Horses see the world through an emotional lens. And just like us, having someone step into our emotional sphere uninvited can be as invasive as a shove.

The Sensory Lead Change Exercise

Creating space means creating comfort and elasticity within the horse's *sensory egg*—that sphere of perceived self-awareness. The horse tells you how big that sphere is and how much emotional space they're occupying, if you're willing to listen.

Approach from each individual sensory zone until you get a subtle signal you've entered it—then back out and move to the next zone. As you move around the horse, you'll notice something nearly imperceptible yet profound: a **Sensory Lead Change** (SLC).

SLC is when the horse successfully "hands off" sensory focus from one zone to the next. We're familiar with physical lead changes—but none are possible without a successful sensory one.

By incorporating this casual, no-pressure exercise into your daily routine, you're speaking the horse's language. It's a gentle warm-up for the sensory system before you warm up the body, and it helps soothe hidden stresses before they displace.

Field Takeaways
- Displacement behaviors signal internal emotional conflict.

- The horse isn't resisting—they're redirecting unprocessed stress.

- Not all stillness is peace; some stillness is shutdown.

- Pawing, excessive chewing, scratching may signal dissonance—not relaxation.

- Emotional redirection can look like "fine" on the outside.

- When you honor silent signals, trust deepens.

- Presence without pressure gives the horse permission to be honest.

Quick Guide: Spotting Displacement vs. Relaxation

Behavior	Relaxation	Possible Displacement
Pawing	Slow, soft, post-release; muscles loosen	Sudden, frequent, in tense context
Lick & Chew	After softening, blinking, lowering head	Rapid, repetitive, no real change in tension
Tail Swishing	Brief, casual, during downtime	Sudden, during pressure or hesitation
Head Movement	Loose, following body rhythm	Swinging or jerking without environmental cause
Focus Shift	Gentle scanning of environment	Abrupt glance away, avoiding engagement
Stillness	Soft-eyed, lowered head, even breathing	Fixed stare, rigid posture, shallow breath

Field Tip: If you're unsure—pause. Notice whether the horse *truly settles* or simply *switches channels* to cope. Relaxation melts tension; displacement just moves it sideways.

Quick Guide: Sensory Lead Change (SLC) Warm-Up

Purpose: To help the horse "hand off" sensory focus smoothly from one zone to the next, creating comfort and elasticity in their sensory egg before physical work.

Steps:

1. **Start Neutral**

   ○ Stand quietly outside the horse's sensory bubble.

   ○ Breathe evenly; let them acknowledge your presence without pressure.

2. **Enter Zone One (Front, Left or Right)**

   ○ Approach slowly from a chosen zone.

   ○ Watch for subtle signs you've entered: ear flick, blink, head turn, softening eye.

3. Back Out

- Step away once you get the signal.
- Give the horse a moment to re-balance before moving on.

4. Move to the Next Zone

- Work around the horse methodically: front, side, hind, opposite side, etc.
- Allow each transition to be led by the horse's awareness.

5. Spot the Sensory Lead Change

- Look for the smooth "hand off" of focus from one zone to another without flinch or freeze.

6. End on Comfort

- Finish in a zone where the horse shows the most relaxation.
- This is your green light to begin physical warm-up.

Field Tip: A successful SLC is invisible to the untrained eye—but to the horse, it's the bridge from emotional readiness to physical willingness.

Trail Marker Sixteen: The Independent Nature

Some Horses Aren't Distant—They're Independent

Not every horse wants to be led. Not every horse seeks closeness. Not every horse gives themselves easily.

And that's not a flaw in the horse. It's a feature of their nature.

Some horses are born with a self-contained emotional system—wired more for independence and less outsourcing. They don't reach outward first. They process inwardly. They may not rush toward you with affection. They wait to see if you've earned the right to be near.

They are not the horses who need you. They are the horses who choose you. And that choice is sacred.

Independence Is Not Rebellion

We often mistake independent behavior patterns for resistance.

- "She's stubborn."

- "He doesn't want to connect."

- "She has no work ethic."

But these horses aren't shutting us out. They're filtering what matters.

They don't follow because you said so. They follow because you *mean* something.

They are the embodiment of discernment—and their loyalty is earned, not owed. For these rare horses in nature, **shared leadership** is everything.

They want to know:

- Are you present, or just pushy?

- Are you aware, or just commanding?

- Are you listening, or just trying to lead?

If you're not attuned to their emotional world, they'll mentally walk away. Not to punish—but to preserve themselves.

Because independent nature in a horse does not tolerate hollowness. They will not follow an empty lead.

Field Note: The Filly Who Didn't Follow

She stood at the far edge of the paddock for three days. Didn't approach. Didn't retreat. She just watched.

Not fearful. Not timid. Just aware.

She was evaluating me—measuring the authenticity of my presence against the pressure I brought with it.

And when I finally stopped *trying* to connect...long periods of silence and stillness—mirroring her patience—she softened.

Horses don't wear watches or carry phones you see...

She walked forward. Not because I invited her—but because I let her decide on her own.

That filly led every other horse in her herd. Not through dominance, but through quiet confidence and presence. She didn't need to prove her strength—because she *knew* it.

And what she taught me was this: Connection isn't claimed. It's chosen, and *presence* will always be chosen over force or urgency.

The Independent Horse Requires Internal Partnership

You don't "train" an independent horse into obedience. You **co-create** a partnership rooted in mutual respect.

- Give them room to think

- Allow space without pressure

- Let pauses be long enough for trust to bloom

- Stop trying to prove yourself—and *be* yourself

These horses don't want a director. They want a partner worth listening to.

They don't bend to control. They respond to clarity.

This is where **you become the bridge**—not the block.

You become the one who reaches across the emotional canyon, not to drag them toward you—but to invite them into a shared space.

You are **bridging the gap between the natural herd dynamic and the domesticated world**—where choice becomes connection, and presence becomes the halter.

Honoring Their Nature Deepens Your Own

When you stop pushing for closeness and start honoring autonomy, everything changes.

You allow:

- Curiosity to replace command

- Silence to speak more than sound

- Distance to become an invitation, not a rejection

You stop chasing trust and start becoming trustworthy.

In doing so, the horse begins to see you—not just as a person in their world—but as a being worth sharing the world with.

That's not training. That's truth.

Field Takeaways

- Independence is not defiance—it's emotional discernment.

- Some horses are wired to observe first, connect later.

- These horses require space, not pressure.

- Trust must be earned—not demanded.

- Co-create a relationship rooted in choice, not compliance.

- Be the bridge, not the block.

- True partnership honors individuality and invites authenticity.

·Quick Guide: Supporting Independent-Natured Horses

Trait	Strength	Potential Challenge	Best Approach
High Self-Reliance	Confident in new or changing environments	May resist unnecessary guidance	Offer partnership, not control; respect their choices
Low Herd Dependency	Can work alone without distress	May appear aloof or disconnected	Build engagement through mutual curiosity
Strong IHD (Individual Herd Dynamic)	Quick, decisive responses to direct stimuli	May miss subtle group cues	Integrate exercises that encourage environmental scanning
Adaptability	Thrives in varied conditions	Can become restless without stimulation	Provide mental challenges and problem-solving tasks
Selective Connection	Bonds deeply once trust is earned	May ignore surface-level interaction	Invest in long-term trust-building over quick compliance

Field Tip: Independent nature isn't isolation—it's self-governance. The goal isn't to "bring them into the herd," but to become the one they *choose* to include in their independent world.

Trail Marker Seventeen: The North Star Never Moves

There Is a Fixed Point in the Emotional Sky

We spend so much time navigating the unknown—training methods, emotional energy, communication, the wild inner maps of both horse and human.

But at the heart of all of it…there is a constant. A point that doesn't change. A truth that doesn't sway.

The North Star of horsemanship is not a method. It's not pressure and release. It's not tradition or trend.

It is relationship. Rooted. Reflective. Real.

And just like the star sailors once used to guide their ships home, this one does not waver—even when the sea is dark.

When in Doubt, Return to the Relationship

Sometimes we all get caught up in:

- Technique

- Timelines

- Goals

- Frustration

We want to do it right. We want to see progress. We want to fix what's "broken."

But the horse doesn't care about perfection. They care about presence.

When the session starts to unravel, when the lesson falls apart, when the connection feels lost—

Look within. Be the North Star never moves.

Legacy Isn't What You Do, It's How You Make Them Feel

At the end of your journey with a horse, they won't remember how clean the transitions were. They won't remember the precision of the bend.

They'll remember:

- The sound of your intention

- The softness of your hand

- The steadiness of your heart

- The feeling of emotional safety

- The space you gave them to be themselves

That is your legacy. Not what you taught—but the thoughtful accountability to their nature.

And in the end, they leave something in us, too.

We Are All on the Trail

This book... this path...this pursuit of discovery—

It was never about instruction It was always about illumination.

To trace the emotional truths that live behind the eyes of the horse. To stand in the quiet spaces where behavior begins. To listen long enough for meaning to rise. To move only when the relationship invites it.

That's the trail. That's the work. That's the journey we share.

The North Star. Why We Follow the Trail.

There are moments in life—quiet, almost imperceptible—when the world slows down just enough for the horse to look back at you.

Not with command. Not with confusion.

But with connection.

In that gaze lives the reason we do this. Not for ribbons. Not for theories. Not even for mastery.

But for **understanding,** for the relationship we have with them helps the relationships we have with others, and ourselves.

To find that rare place where how they feel and how we show up exist in harmony.

This Is the Compass

Every observation in this book, every map, term, trail marker, and insight—isn't just about the horse.

It's about **how we choose to see the horse.**

This is the North Star.

To live the work means to follow that star—even when it leads us away from what's convenient. Even when it challenges what we thought we knew.

Because the journey isn't linear. It's emotional. And the greatest compass we'll ever carry is the willingness to walk **beside** the horse, not in front of them.

We don't teach answers here. We illuminate patterns.

We don't break horses down. We build trust upward.

And in this pursuit of discovery, what matters most is not that you reach the end—but that you keep following the trail with an open mind and an even softer heart and hand.

Because somewhere along the way, the horse will show you something you didn't even know you were looking for.

And in that moment...**you'll know you're home.**

Thank You~
Kerry

GLOSSARY OF TERMS

A Field Lexicon of Emotional Understanding

This isn't just a list of definitions—it's a window into the language behind the language.

These terms form the heartbeat of Herd Dynamic Profiling™, shaped by years of field observation, emotional mapping, and lived experience. Each one carries with it not just meaning—but memory.

Anticipatory Response Mechanism (ARM)

The horse's predictive processing system—rooted in memory and expectation. Rather than reacting to what *is* happening, the ARM prepares the horse for what they believe *will* happen, based on past emotional patterns and associations.

It can be helpful in familiar environments—but when shaped by fear or trauma, it can create automatic reactions to ghosts of moments past.

Understanding the ARM allows us to disrupt emotional assumptions—and replace them with trust-building presence.

Associative Aspect

The horse's ability to emotionally link one stimulus, environment, or experience with another.

This is why horses may react to seemingly unrelated cues with eerily similar behaviors—because to them, it *feels* the same.

The trailer isn't just a trailer—it's the place they left their friend. The halter isn't just a halter—it's the thing that always comes before stress.

Understanding the associative aspect helps us trace not just behavior—but what that behavior has come to *represent*.

Compression

The buildup of internal pressure when the horse experiences more stimulus than they can process, filter, or release.

Compression stress often precedes blow-ups, shutdowns, or emotional retreat. It may not be visible—until it manifests as behavior.

Collision

An emotional or sensory overload event—when the horse's cushion fails and the nervous system spirals into flight, freeze, or reactive protection.

Collisions are not defiance. They are signals that the emotional pathways have been overwhelmed.

Cushion

The emotional buffer zone between stimulus and reaction. The emotional "workspace".

A healthy cushion lets a horse absorb the world without being consumed by it. When the cushion is strong, they can think. When it's thin, they can only survive.

We don't train the cushion. We protect it and help expand it, make it elastic under stress.

Emotional Energy

The internal fuel that powers response, communication, and connection.

Emotional energy can be:

- **Expressed** (through action or interaction)

- **Suppressed** (internally stored or hidden)

- **Displaced** (redirected into unrelated behavior)

Learning to read emotional energy tells us where the truth of behavior lives.

Emotional Response Mechanism (ERM)

The horse's psychological system for interpreting and responding to emotional stimuli.

It shapes:

- Speed of reaction

- Depth of focus

- Authenticity of behavior

The ERM operates through either the IHD (individual focus) or GHD (group dynamic)—and it's where pressure becomes either opportunity... or overload. Closely related to but not eh same as, ARM.

Group Herd Dynamic (GHD)

The horse's capacity to multitask, interpret multiple stimuli, and "read the room" within a social or environmental context.

It's how a horse maintains awareness of herd, surroundings, and subtle energetic cues—all at once.

A strong GHD supports adaptability, calmness in chaos, and collective cohesion. The underwriter of expressed athleticism.

Individual Herd Dynamic (IHD)

The one-on-one emotional focus system - *expressed athleticism*.

IHD governs how a horse bonds with a specific person, horse, or task. It's the channel of singular attention and emotional targeting—the essence of connection, both physical and felt.

Independent Nature

An internal trait marked by emotional self-reliance and inward processing. These horses don't follow by default. They assess. They wait. And when they come to you—it's by choice.

Not aloof. Not stubborn. Independent. They don't want control. They respond only to shared leadership.

Sensory Lead Change

A moment of internal redirection—when the horse switches which sensory channel is leading the experience.

Like a physical lead change in motion, this is a psychological pivot.

It may show up as a pause, shift in attention, or momentary stillness. Recognizing it helps you meet the horse where they're *going*—not where they've been.

Sensory Soundness™

A term coined to describe a horse's emotional and environmental clarity—their ability to accurately interpret, filter, and respond to incoming stimulus.

A sensory sound horse isn't just physically capable. They're emotionally calibrated.

This is the foundation for trust, learning, recovery... and performance.

Stimulus Interpretation

The process by which the horse assigns meaning to the world.

Two horses may encounter the same object—but interpret it entirely differently based on their emotional filters, memory, and herd dynamic wiring.

How they see the world **is** how they respond within it.

Tactile Awareness

The horse's sensory awareness through touch, proximity, and physical feedback.

It's how they feel:

- Pressure

- Presence

- Subtle changes in space

Tactile awareness isn't just about safety—it's how horses bond, explore, and connect with others... and with you.

Threshold Moment

A defining point in time when a horse either connects or disconnects. This is the emotional fork in the road— where fear can take root or trust can rise.

Handled with patience, these moments become breakthroughs. Missed or forced, they become blockages. Threshold moments are not about technique, but rather about stress management.* They are about timing, feeling, and presence.

A Note From The Field

"A literary field companion into the heart of the horse... and the soul of connection."

Kerry's work has completely transformed how I understand and work with my horses. Through Sensory Soundness™, I've come to deeply appreciate how each horse processes their environment and how that impacts their behavior, performance, and relationships—with both other horses and their human partners.

One of the biggest shifts in my approach has been understanding the concept of "outsourcing." Knowing that a horse can look to me as their external processing center has helped me become a more consistent, calming presence in their world. This insight has influenced not only how I train my own horses, but also how I coach young riders. It's been especially powerful when working through common issues like spookiness—where understanding a horse's sensory threshold can make all the difference.

With my mare Clover, we developed a system that helped her process her environment more effectively and stay mentally present with her body. Before, her mind would often get too far ahead of where she was physically, which impacted her performance and focus. With small, tar-

geted training aids and a better understanding of her sensory processing style, we saw a noticeable improvement in her outward performance at competitions.

Kerry's work has brought a deeper level of empathy and effectiveness to both my riding and my teaching. This book is a must-read for anyone who wants to build a stronger, more intuitive partnership with their horse.

Tegan Lush - 5* Eventer, Australia
Tegan Lush Eventing Team
www.orangeandblack.com.au

About the Author

Kerry M. Thomas is an equine behavioral analyst and the founder of **Herd Dynamic Profiling™** and **Sensory Soundness™**—internationally recognized frameworks for understanding how horses think, feel, and emotionally process the world around them.

For over two decades, Kerry has worked with horses in both wild and domestic settings—from high-performance racehorses to trauma survivors—with a singular mission: to uncover the emotional language that shapes behavior and defines trust.

His work blends instinct, research, and reverence for the unseen. He lives in Pennsylvania with his wife, **Daphne**, where together they continue to study, teach, and walk gently in the company of horses.

Presentations and workshops available
Visit: www.kerrymthomas.com

www.ingramcontent.com/pod-product-compliance
Lightning Source LLC
Chambersburg PA
CBHW060551030426
42337CB00021B/4537